# PICTURING AFGHANISTAN

## The Photography of Foreign Conflict

THE HAMPTON PRESS COMMUNICATION SERIES
Communication, Globalization and Cultural Identity
Jan Servaes, series editor

The Dao of the Press: A Humanocentric Theory
*Shelton A. Gunaratne*

Connectivity, Networks, and Flows: Conceptualizing Contemporary Communications
*Andreas Hepp, Friedrich Krotz, Shaun Moores & Carsten Winter* (eds.)

Viewing the South: How Globalization and Western Television Distort Representations of the Developing World
*Emma Miller*

Cultural Citizenship and the Challenges of Globalization
*Wenche Ommundsen, Michael Leach & Andrew Vandenberg* (eds.)

Picturing Afghanistan: The Photography of Foreign Conflict
*Paul Verschueren*

Serbian Spaces of Identity: Narratives of Belonging by the Last "Yugo" Generation
*Zola Volcic*

Media and Power on the Margins of Europe: The Public Negotiation of the Breton Language and Cultural Identity
*David P. Winterstein*

forthcoming

New Voices Over the Air: The Transformation of South African Broadcasting in a Changing South Africa
*Eric Louw and Viola Milton*

# PICTURING AFGHANISTAN

## The Photography of Foreign Conflict

Paul Verschueren

HAMPTON PRESS, INC.
NEW YORK, NEW YORK

Printed in the United States of America

Library of Congress Cataloging-in-Publication

Verschueren, Paul.
Picturing Afghanistan : the photography of foreign conflict / Paul Verschueren.
p. cm. -- (The Hampton Press communication series. Communication, globalization and cultural identity)
Includes bibliographical references and indexes.
ISBN 978-1-61289-038-8 (hardbound) -- ISBN 978-1-61289-039-5 (paperbound)
1. Afghan War, 2001---Photography. 2. Afghan War, 2001---Pictorial works. 3. Afghan War, 2001---Press coverage. 4. Afghan War, 2001---Political aspects. 5. Psychological warfare--Afghanistan. 6. Propaganda--Western countries. 7. Public relations and politics--Western countries. 8. Journalism--Political aspects--Western countries. 9. Photobooks--Political aspects--Western countries. 10. Mass media and war--Western countries. I. Title.
DS371.4135.V47 2012
070.4'499581047--dc23

2011050234

Hampton Press, Inc.
307 Seventh Avenue
New York, NY 10001

To Chris

# CONTENTS

## List of Illustrations

# PREFACE

This book explores dominant Euro-American ideas about the conflict in Afghanistan through an examination of the use of photography in four domains: public affairs, psychological operations (PSYOP), the press, and the photo-book. These areas usually are considered separately: Photo-books are an area of study in art history; communication scholars often focus on the press; public affairs attract the interest of public relations researchers; and PSYOP are a subject in military studies. By looking at images and their captions across these domains, I demonstrate that audiences at home and in the war zone have been targeted with similar messages about the nature of the conflict, and that there are consequently no strong boundaries between photography used in warfare and photography about war. I also show when and how public affairs uses the aesthetic codes of the photojournalistic genre to suggest a reading of photographs as morally neutral and objective, when and how press photography borrows concepts, themes, and aesthetic strategies from public affairs, and when and how photo-books deviate from the dominant official and media discourses.

Relations between public affairs, PSYOP, the press, and the book can be studied from various angles, but this work focuses on the role of photography. In the 1990s, when digital technology began to rise, some believed that photography no longer mattered and would become obsolete. Now most observers agree that the reverse has really happened (Lister, 2007). It now seems that photography, the foundation of modern vision, still is central to multiple contemporary forms of communication and culture. In diverse ways, and through more channels than ever before, people continue to use photography as a central medium to express their relationship to the world and to others. Today, multiple media consider the photographic indispensable, not just to inform or entertain audiences, but to position themselves on the market as well. And although people may doubt the veracity of photography today, with its emotional impact and naturalistic qualities, it remains an important rhetorical and political device. All modern wars are "wars of images," and photography seems a key factor in the use of culture to exercise influence, not in the least because it always and inevitably suggests a role for onlookers to play. But there is no straightforward way to assess the workings of photography as a soft-power factor. Photographs are capable of supporting and questioning war, but they rarely do so on their own. As many have stressed, pictures are usually surrounded by words that attribute meaning to them (e.g., Barthes, 1977; Hall, 1981; Scott, 1999; Sontag, 1979). In accordance with this view, I do not simply treat the photograph as a separate realm but as a space framed by texts, and in particular by the caption. Taking this notion of the textual frame further, and in accordance with developments in photo theory since the 1980s (e.g., Bolton, 1989), I also consider the photograph as an object framed by the institutions that use it, acknowledging that the meaning of a photograph is largely located outside itself, in its physical location, its accompanying texts, and its institutional settings.

My interpretive approach is based, in a general way, on insights from the three genealogies that Elkins (2003) suggested for visual studies: art history, cultural studies, and semiotics and communication theory. In art history, Arnheim (1954) and Gombrich (1960) shifted attention from the individual artist to socially shared ways of seeing, and to perception as a structuring process. Cultural studies extended this perspective by drawing attention to the role of visual images as co-determinants of unequal social relationships. Common ways of seeing were approached as sociocultural patterns linked to positions of power. Although I generally accept these views, most of my inspiration for this work has come from communication theory and semiotics, visual studies' third genealogy that is, unlike cultural studies, organized around theoretical objects rather than political ideas (Mitchell, 1995).

From communication theory I borrowed the concept of framing and from semiotics the tools to identify frames and associated cultural values. Framing draws attention to the role of selection and focus in the creation of meaning. Professional photographers usually are well aware that they suggest meanings through these mechanisms, and many seem to believe that their framing choices can influence public attribution of responsibility and acceptance or rejection of possible solutions to a problem. Using the concept of the frame, I show how certain framing choices support certain "truths" about conflict, and I investigate some of their moral and political implications. The second chapter in this volume explains this in more detail as it tries to establish a bridge between framing and social semiotics. Let me just say here that my perspective is informed by an anthropological understanding of communication as a process determined by time- and space-specific beliefs and practices (Carey, 1989) and by the now well-established constructivist perspective

(P. L. Berger & Luckmann, 1966). Photographs are interpreted in accordance with this paradigm as configurations of the references that their producers found important. Within conventionalized fields such as photojournalism, their references usually reflect and sustain shared and commonsense perspectives on the world and others. As I show here, outside these areas, but occasionally within them too, photography also operates as a cultural ritual that creates ambivalence, hybridity, and liminality. Without attention to the unstable and ambivalent aspects of representation, one can easily jump to distorted and homogenized conclusions about Western representations of Afghanistan. Although this work groups, at the risk of such a distortion, a multiplicity of meanings into a relatively small number of general categories, called "*master* frames," I have tried to be attentive to the disorderly, multivarious, ambiguous, and contradictory aspects of photography as much as possible too.

Because all interpretations of photography are dependent on specific acts of looking, they are always provisional and partial. The interpretation of photography is a particularly slippery matter because its syntax is implicit; its artificial perspective equated with natural human vision; and, in many contexts, the ideology of its practitioners is softened by professional and commercial demands. By being as precise as possible about my own research frame (i.e., my research questions and concepts), I tried to create a reasonable and reliable account of the role of photography in the representation of the conflict in Afghanistan, and one that will contribute to a better understanding of the relations between the visual, the violent, and the foreign. I have grounded the analysis as firmly as possible in the systematic observation of images and their context. The analysis is based on photographs and their surrounding texts, not on isolated visual images. Information about their cultural, economic, social, political, and military contexts is included, because photographers and writers work within the constraints of these environments. But although I pay some attention to the local circumstances in which photographs were made, this book is in no way an attempt to reconstruct the social behavior and background of their makers. The focus is on the images and their texts, how they are grounded in master frames, and how these frames converge and sometimes collapse.

Although this book was finished when U.S. President Barack Obama began to counter-frame the "war on terror" rhetoric of his predecessor, I believe the present analysis remains important because the frames it describes do not merely belong to the Bush era but are rooted in Western beliefs that often date back to colonial times. I am aware that this work has many limitations and that much more remains to be examined. Let me mention two areas that seem important to me. First, the history of the Western visualization of conflict in Afghanistan has not yet been written. In particular, the photographs of the Soviet–Afghan War (1979-1989) are waiting to be rediscovered. I wish I could have included an account of these images, and of the colonial imagery at the time of the Great Game (1800-1844) too, when the Russian and British empires competed for influence in the region. But that would have postponed the publication of this book further. That work remains to be done, and might shed more light on continuities and changes in framing conflict in Afghanistan. Second, the role of photojournalism in contemporary popular culture merits more attention. I devote some attention to the contexts in which photojournalists operate today, but a lot more can be said about the current and changing conditions of journalistic production, both in and outside the context of war reporting. I hope that despite these and no doubt other limitations the present work will prove useful and will offer some new insights and explanations.

# ACKNOWLEDGEMENTS

Several people helped me through the creation process of this study. My research started as a PhD project conducted with support from the School of Journalism and Communication at The University of Queensland, Australia. Jan Servaes later suggested this book and provided me with the opportunity and feedback to formulate my thoughts more clearly and to develop my ideas further. I sincerely thank him. I also am grateful to René Devisch and Leo Van Audenhove, who kindly supported my original research application, to the staff at the School of Journalism and Communications at The University of Queensland, to the anonymous reviewers of my work, and to my publisher. Multiple sources have influenced my writing and thinking, and my interest in photography. Although not cited on any of the following pages, David Levi Strauss' *Between the Eyes* (2003) inspired me to investigate relations between the photographic and the political, and I also owe much to Caroline Brothers' *War and Photography* (1997), which sparked my interest in conflict and photography further. There are numerous photographers, with diverse interests, styles and practices, who have stimulated my interest in photography, including Philip Blenkinsop, Carl De Keyzer, An-My Lê, Gilles Peress, Susan Meiselas, Richard Mosse, Sophie Ristelhueber, Lisa Ross, Paul Seawright, Fazal Sheikh, Guy Tillim, and many more. Finally, I thank my wife Chris and my children Thanit and Thanya for their love and patience while I was writing this book. Without the support of Chris, I probably would never have been able to finish it.

# Introduction

> See, there is only a single character. It is the barbarian character *war*, but it has other senses too. It can stand for *vengeance*, and, if you turn it upside down like this, it can be made to read *justice*. There is no knowing which sense is intended.
>
> —J.M. Coetzee (1982)

At the beginning of Coetzee's *Waiting for the Barbarians* (1982), the protagonist of the story is convinced of his qualifications to speak about the barbarians. A magistrate in the service of the empire, he later looses faith in his employer and becomes uncertain about his own role and identity. The magistrate no longer understands what war is for, who the barbarians are, who he is, and to whom he belongs. Similarly, post-9/11 photo-textual representations of Afghanistan are located between the self-confident and the ambivalent. When airplanes hit the World Trade Center in New York and the Pentagon in Washington on September 11, 2001, the West condemned the act quasi-unanimously as barbaric. But plans by the United States to retaliate in a military way, and the actual bombing of Afghanistan, made differences between "the empire" and "the barbarians" less evidently clear within weeks after 9/11. By the end of November 2001, British journalist Robert Fisk (2001) concluded in the *Independent* that a complete reversal of roles had taken place. Responding to the violent killing of Taliban prisoners in Mazar-e Sharif, Fisk entitled his angry article "We Are the War Criminals Now," whereas not much earlier Philip Hensher (2001) had still suggested in the same newspaper with confidence that "the ordinary Afghan would benefit to a colossal degree from the imposition of our cultural, political and even religious values" (para 12). After Fisk's article, politicians and military leaders continued to express their belief in a Western presence in Afghanistan. Gen. Tommy Franks, former commander-in-chief of U.S. Central Command, confidently called the intervention in Afghanistan "the most precise war in America's history, to be sure, beyond question" (Grier, 2002). Over the years, claims about the intervention in Afghanistan have become ever-more modest, and initially strong convictions have been replaced by more vague assertions that *success is still possible* and *NATO cannot afford to loose in Afghanistan*. Eight years after the first arrival of Western troops in Afghanistan, even Robert Gates, U.S. defense secretary and U.S. commander of NATO, expressed his doubts about the Western presence in Afghanistan and made a reversal analogical to that of Fisk: "My worry is that the Afghans come to see us as part of the problem, rather than as part of the solution" (cited in Barry & Thomas, 2009, p. 18).

As this book intends to show, photo-textual representations[1] of Afghanistan since 2001 can be located between the two opposing trends: the first toward cohesion and the confident confirmation of existing power relations, the second toward more uncertain and ambivalent positions and reversals. Since 2001, photographs of Afghanistan have served, in straightforward ways, the reification of the national community and the reproduction of power relations but they also have reflected and construed, at least sometimes, the fear of the breakdown of social structures and identities.

Especially at times of war, the celebration of the imagined national community (Anderson, 1983), in order to instill confidence in the public, often happens through the creation of an us-and-them dichotomy that, within the context of encounters with non-Western peoples, usually is established through large generalizations as described by Saïd in *Orientalism* (1978). Saïd defined Orientalism as a relationship of power that is reproduced by discursively positioning the Orient as an exotic object of Occidental study and by contrasting positive representations of the Occidental self with negative representations of the Oriental other. Many contemporary photo-textual representations of Afghanistan are illustrative of the exoticism and binary schematization of Orientalism, but the conflict in Afghanistan has also stimulated a "terror of uncertain signs" (Barthes, 1980, p. 247). Representations of Afghanistan sometimes have been more complex, unstable, and uncertain than analyses in terms of us-and-them or the exotic suggest.

Because of its lack of formal syntax, photography—even when controlled by the hierarchy of the syntax of captions and headlines—often promotes positions of uncertainty and ambivalence. One of my arguments in this book is that, despite tendencies to celebrate social integration through reinforcing cultural differences, ambivalences have been a part of the photo-textual representations of Afghanistan, at least in western Europe, from early October 2001 onward, when U.S. military began to bomb the country. Substantial parts of the political, military, journalistic, and humanitarian fields have received the conflict in Afghanistan with doubt since 2001, and photo-texts in the press have regularly reproduced this hesitation about the purpose and nature of the intervention. At the beginning of the conflict, the ambivalent accounts mainly revolved around the effectiveness of the military strategy, the loss of control, the inability to maintain borders, the fear of global disorder, and the convergence and obliteration of identities. At least in western European media, the intervention in Afghanistan appeared from its very beginnings as liberation and oppression, as death and resurrection, as abstract and concrete, as local and global, as territorial and de-territorial, as vengeance and justice, as constructive and destructive. Often there has been, in the words of Coetzee (1982), "no knowing which sense is intended" (p. 112).

In immediate post-9/11 conflict reporting, the myths of national collectivity and cohesiveness never were fully maintained. It appears that even within highly conventionalized genres such as photojournalism, images are capable of creating an indeterminate space of signification that disrupts conventional categorization. Photography not only is a medium of social integration (Bourdieu, 1990) but of discontinuity, in the sense of the liminal (Turner, 1967), the hybrid (Bakhtin, 1981; Bhabha, 1994), and other terms that focus attention on disruption and difference, such as inversion (Babcock, 1973), abjection (Kristeva, 1982), and heterotopia (Foucault, 1984). These concepts, each with its own emphasis and analytical scope, suggest a failure of order, a dissolution of established borders, and possibly a subversion or reinvention of the conventional. They draw attention to photography beyond its role as a ritual supporting social unity.

Hybridity originally was defined by Bakhtin (1981) as "a mixture of two social languages within the limits of a single utterance, an encounter, within the arena of an utterance, between two different linguistic consciousnesses, separated from one another by an epoch, by social differentiation, or by some other factor" (p. 358). Since Bakhtin, the term has been used to describe a quasi-permanent tension of social reality on the one hand and a temporary liminal space between two stable zones or identities on the other hand (Rosaldo, 1995). The liminal (Lat. *limen*, i.e., *threshold*), Turner's concept (1967), stressed the marginal, ambiguous, and unfamiliar aspects of the transitional stage in a rite of passage. Turner described this stage as in between two other stages, which he called separation and reincorporation. He interpreted liminality in traditional societies as the moral demand to respect the established order, although he also described liminal situations as opportunities for renewal and "seedbeds of cultural creativity" (Turner, 1982, p. 28). The latter aspect became more prominent in his account of the liminoid (Turner, 1974), a term that he introduced to describe activities of play, rejuvenation, and subversion during leisure time in modern societies,[2] and in later accounts of hybridity (Bhabha, 1994). Although Turner's theory of the liminal suggests that ambivalences indirectly support the status quo, the poststructuralist and postcolonial understandings of hybridity imply—in line with Turner's later concept of the liminoid—that ambivalences, contradictions, and transgressions are an integral part of social reality. The coexistence of contradictory forces is no longer seen as an aberration of the normal but as a quasi-permanent characteristic arising from differences in social experience. The focus on the antistructural in liminal theory is replaced by the question of contingency and tradition. Bhabha and others conceptualized contingencies as opportunities for cultural renewal and change. Applying the idea of creative renewal to the issue of identity in the context of colonialism and postcolonialism, hybridity was defined as a positive value against interpretations of the world in binary terms: West–East, self–other, secular–religious, modern–traditional, masculine–feminine, adult–child, and so on. Locating the source of hybridity within language itself, as Bahktin (1981) did, Bhabha understood the indeterminacy of the sign not as a mere "terror"—a constant and threatening deferral of meaning—but as an opportunity for counter-hegemonic emancipation.

The photographic sign is indeterminate and potentially subversive. The image known on the Internet as "Mullah Bush" (Radiant World, 2001), reproduced on the cover of *The Clash of Fundamentalisms* (Ali, 2002), illustrates how photography is able to dissolve established binary positions. The digital montage merges a portrait of U.S. President George W. Bush with one of a man who looks like a Taliban Mullah (*religious leader*) and Al-Qaeda leader Osama bin Laden. The composite creates a new space of meaning that invites the viewer to rethink the two entities. The interstitial space is subversive of both sides: It questions the politics of the Bush administration and that of the Taliban regime at the same time. Hence, it suggests the breakdown of established political orders in a more general sense.

Other images have exploited photography's hybrid capabilities and ambivalent qualities in similar ways. The Russian photo-conceptualist group AES, now called AES+F, produced a digital image of the Statue of Liberty clad in a burqa or *chadri*, which it labeled "New Liberty." The image is part of the AES+F's *Islamic Project* (1996-) in which various Western iconic spaces are hybridized with symbols related to Western perceptions of Islam. Whereas "Mullah Bush" primarily seems to suggest the clash and breakdown of *political* order, "New Liberty" draws attention to the possible collision and dissolution of established *cultural* orders. Probably for that reason, it was used on the cover of a volume called *Culture and Conflict in Global Perspective* (Bertelsmann Stiftung, 2010).

Similar examples of cultural hybridization include various pictures of the American Stars and Stripes used as a burqa, and a photograph of rapster Lil'Kim in a burqa-like dress on the cover of the American fashion and music magazine *One World* (Complex, 2003). Other composite images destabilize established *economic* relations. Antonio Mendoza of www.subculture.com, for example, linked an image from a bin Laden video to an old Apple Computer advertising slogan, "Think different," associating Islamist terrorism with U.S. capitalism and technology, and questioning the meaning of both.

The examples suggest that representations of Afghanistan cannot simply be reduced to self-confident projections of power with clear demarcation lines between the East and the West. In a global world, the borders of an imaginary East–West geography are difficult to maintain, and the state of uncertainty and anxiety that results from it has been reproduced, as I hope to show, in press photographs and in some of their accompanying texts, too. For, although journalists generally strive toward unified narration, ambivalences certainly are not absent from the news or limited to artistic composite photography. For example, the mainstream press portrayed bin Laden, not unlike Antonio Mendoza, as a capitalist with properties around the world, dressed in an American camouflage jacket, and with a wristwatch made in the United States. Bin Laden was represented as neither fully here nor there, and therefore dangerous.

As I show in Chapters 2 through 4, Afghan refugees in Europe, "American Taliban" John Walker Lindh, and European Muslims were portrayed in a similar fashion, as liminal entities that challenged the boundaries between simplified, stereotypical Eastern and Western identities.

Three general questions guided the analysis:

1. In what ways has the conflict in Afghanistan been framed?
2. Has the conflict in Afghanistan been presented from an international, domesticated, or Afghan perspective in the western European press?
3. What is the relation between official U.S. frames and western European media frames?

The first research question concerns the framing of conflict and of national and international relations. It is inspired by the central thesis of peace journalism,[3] which holds that mainstream media predominantly frame conflicts in a war-oriented rather than peace-oriented manner. In Galtung's work (e.g., 2002), peace and war journalism are seen as opposing frames construed around four central notions: peace–conflict versus war–violence, truth[4] versus propaganda, people versus elites, and solutions versus victory. War journalism dramatizes and domesticates conflicts; peace journalism humanizes and localizes them. Challenging traditional war journalism's tendency to focus on battles, victories, technologies, elite leaders, troops, and material damage, peace journalism is a solutions-oriented way of reporting that explores structural inequalities, the small dramas of local people, and their peace initiatives.

The second research question about the domestication of news relates to one aspect of war journalism and, more generally, to explanations of differences in reporting across nations. Researchers found that journalists cover foreign conflicts from a national perspective, especially when their own country is involved (Entman, 1991; Wolfsfeld, 1997). When foreign frames are not supported by domestic organizations, it is unlikely that they will enter the media discourse (Wolfsfeld, 1997). Other research confirmed that international news is made to fit the expectations and understandings of the public at home.[5] For example, Swanson and Smith (1993), comparing television news reports on the first Gulf War in seven countries, found that "there was no global television news story of the war; there were seven different village stories" (p. 188). Gurevitch, Levy, and Roeh (1991) interpreted domestication as a counterforce to globalization and as a means to protect national identity. Other authors, from a different perspective, similarly argued that globalization may well be strengthening rather than weakening national perspectives (Castells, 1997a, 1997b; Featherstone, 1995). Other studies stressed the homogenization of news due to new technological capabilities, the internationalization of news agencies, media convergences and international standards in journalistic practice (e.g., Boyd-Barrett & Rantanen, 1998; Josephi, 2005). Bringing both perspectives together, news production can be seen as determined by both globalization and domestication pressures.

The third research question concerns divergent views about the role of the West in Afghanistan since 2001. Shortly after the 9/11 attacks, and before the intervention by the United States, Europeans already responded nervously to the rhetoric of President Bush and his administration. The president's reference to a crusade against terrorism raised fears of a clash of civilizations and the blunt exercise of military power. The military logic of the "war on terror" concept of the Bush administration was translated into a focus of the United States on combat and counter-insurgency missions under Operation Enduring Freedom, while European nations generally engaged in peacekeeping-related activities under the International Security Assistance Force, first a European UN-mandated organization, later a NATO-led mission. But perhaps differences across the Atlantic—between a so-called soft-power Europe and hard-power United States—should not be overstated. In the course of time, there has been considerable rapprochement between Operation Enduring Freedom and the International Security Assistance Force. Also, public opinion research indicated that Europeans perceive the world largely as Americans do. A Worldviews survey (2002) concluded that Europeans appeared critical of U.S. foreign policy but were nevertheless willing to use force. An emphasis on divergences across the Atlantic may also obscure internal divisions. Countries and political parties in Europe have certainly been internally divided on their role in Afghanistan. In 2001, Germany's green party was divided over the country's participation in the allied intervention, and 9 years later the Dutch government collapsed over Afghanistan.

Overall, divergences within Europe and across the Atlantic were much greater with regard to the war in Iraq, when U.S. Secretary of Defense Donald Rumsfeld labeled countries that were critical of the war—Germany and France in particular—as "Old Europe" (BBC News—World Edition, 2003). The comparison that I make between official U.S. views and western European press views in this book suggests that western European reports about the liberation of Afghanistan from the Taliban and Al-Qaeda often were ambivalent. The western European press was critical of the military strategy but generally not of the use of force. Afghanistan has usually been framed in western Europe as former U.K. Shadow Home Secretary David Davis (2009) did in a piece for the *Independent* entitled "The Right War Fought in the Wrong Way."

To explore the three research questions in a reasonably broad way, I collected various sets of data. The first set, discussed in Chapter 2, concerns the official U.S. view on Afghanistan. Official views typically are derived from verbal statements such as government policy documents, commentaries, and speeches. In this work, I focus on the visual images that operate in conjunction with these and that may play a role in supporting or contesting official ideas. Because of their indexical and iconic qualities, images can provide convincing "evidence" of what goes on, and they also may have a more direct and emotional appeal than texts (Paivio, 1986). Combat images in particular seem to elicit an emotional response, which may influence viewer attitudes about a military operation more effectively than texts alone (Pfau et al., 2008).

Because U.S. policy on Afghanistan made a distinction from the start between a military operation aimed at the Taliban and Al-Qaeda, and a humanitarian operation aimed at the local population, I look at photo-textual representations by two U.S. organizations that have embodied this approach: the Department of Defense (DoD) and the United States Agency for International Development (USAID).[6] The DoD images that I discuss include military public affairs photographs intended for home audiences as well as psychological operations (PSYOP) images intended for use in Afghanistan.[7] Because of the high degree of illiteracy in Afghanistan, photos and illustrations have played a central role in PSYOP.[8] It could be argued that the public affairs photographs are images *about* war, whereas the PSYOP images are used *in* war, and that both are therefore unrelated. To the extent that the conflict in Afghanistan has global dimensions and home audiences have been implicated in the "war on terror," this is not true, however. PSYOP and public affairs are both forms of propaganda used to influence civilian attitudes and perceptions. From the beginning of the intervention onward, home audiences *and* Afghan audiences alike had to be convinced of the evil of Al-Qaeda and the Taliban, and national unity had to be strengthened both at home and in Afghanistan. As I hope to show, this resulted in the use of similar imagery abroad and at home.

The second set of images, discussed in Chapter 3, was taken from eight western European elite newspapers with a higher educated readership of political, economic, and cultural decision makers, managers, and professionals, and those aspiring to such positions. The choice of newspapers was not, in the first place, guided by how many readers these newspapers attract. The newspapers were selected to get a fairly broad overview of the elite perspectives in western Europe. For this reason, I included newspapers from four countries, in four languages, with different political orientations, and different approaches to news and photography. Of the newspapers, 50% are generally regarded as left or left-centric (*Guardian, Le Monde, Die Zeit, De Morgen*), the other half are more conservative (*Times, Le Figaro, Die Welt, De Standaard*). All newspapers are dailies, except for *Die Zeit*. *Die Zeit* is a weekly newspaper but was included because it is currently Germany's most widely read elite newspaper, and the weekly political press in Germany occupies a more prominent place than in the other countries.

I examined all photographs related to Afghanistan that were published in these newspapers between September 11, 2001 and February 28, 2002. The photographs had to be made in Afghanistan or had to have a reference to Afghanistan in their caption in order to be included in my sample. After December 22, 2001, when the Interim Authority in Afghanistan had been installed, the number of photographs in the press diminished rapidly. I collected and analyzed all photographs from the next 2 months, but these images did not reveal any new results. I decided at that point that the data had been exhausted, and that more data would no longer substantially improve the analysis and overall quality of the research.

Images from the U.K. newspapers may appear to be overrepresented in the discussion. There is an obvious reason for this. The United

Kingdom has been more involved in the intervention, and Afghanistan belongs to its historical sphere of influence. As a result, the U.K. newspapers have covered the conflict in Afghanistan more extensively. They have published more photographs, more articles, and have sent more journalists to Afghanistan.

All the newspapers display singular photos and rarely present an elaborate sequence of images as is more customary in magazines. Popular papers were not included, but an informal look at 100 front pages of the U.K. newspaper *Mirror* seems to suggest that popular and elite newspapers, when they are situated at the same end of the political spectrum, are likely to offer similar views on foreign conflict. What differs substantially is their presentation of imagery. Popular newspapers are more explicit in framing reality than elite newspapers. Points of view in elite newspapers are usually more implicit and more difficult to observe.

The third and final set concerns a small number of photo-books. Ron Haviv's *Afghanistan: The Road to Kabul* (2002a), Simon Norfolk's *Afghanistan: Chronotopia* (2002), Paul Seawright's *Hidden* (2003), and Thomas Dworzak's *Taliban* (2003) are discussed in Chapter 4. These books do not, like the newspaper data, correspond to what most decision makers or audience members are likely to see, and are not, like the PSYOP leaflets, targeted at the local population. They reach only a small audience but are included to offer an insight into alternative narratives outside the dominant official and media frames. With the exception of Ron Haviv's work, the books are framed to varying degrees in opposition to popular media accounts. They can be qualified as documentary in the general sense of not being concerned with the immediate, dramatic or sensational that often preoccupies the photojournalist.

A focus on photography in studying Western representations of Afghanistan seems justified for several reasons. To begin with, it has been said that Western epistemology is to a large extent ocular centric (Zelizer, 2001). The scopic regime of modernity privileged seeing as a major way to achieve objective knowledge (Metz, 1977). Photography gained its status within this context from its presumed unique ontological feature, its indexicality.[9] But debates about the nature and effects of photography remain unresolved. There is a substantial body of literature that suggests important effects for photography, but other research indicates that these effects are mediated to a considerable degree by other forces. Studies in advertising found that images are influential devices (Beasley & Danesi, 2002; Messaris, 1997; Messaris & Abraham, 2001). An experimental study by Barrett and Barrington (2005) suggested that newspaper photographs can influence the evaluation and impression of a political candidate and the decision to vote or not to vote in favor. The power of photography appears to be especially great when people have no direct experience of the depicted reality, as in the case of foreign news. The tendency to believe what we see seems to have a neurophysiologic base. When the human brain registers something through the eyes, it seems to consider it as if it physically exists (Newton, 2001).

Although texts constitute a key determinant of photographic meaning, the processing of verbal information is often guided by visual information as well. Images attract the attention of readers and serve as points of entry to texts (Barnhurst, 1994; Mendelson & Thorson, 2004). A study based on research with eye-tracking equipment confirmed that readers look at pictures and strong graphics first, then read the picture caption, then the headline, and then the story (Garcia & Stark, 1991). Articles accompanied by photographs—especially photographs depicting victimization—are read for longer periods of time (Zillmann, Knobloch, & Yu, 2001). Other researchers found that photographs operate in less obtrusive ways than texts but may convey more controversial meanings (Messaris & Abraham, 2001). More generally, visual images, because of their high degree of iconicity, often add affect and emotion to messages (Peters, 2005). Many argued that still images are more memorable than moving images (Evans, 1978; Sontag, 1979), and studies in psychology confirmed that messages with a visual component are more easily understood and retrieved from memory (Lynn, Shavitt, & Ostrom, 1985). It is further believed that visual images matter because we are flooded with them. Sontag (1979) and Moeller (1999) argued that the ubiquity of photography numbs us down. Sontag insisted that we can no longer be moved by images because we have seen too many, and Moeller, along similar lines, suggested that the sheer flood of images of misery and conflict has led to "compassion fatigue." Both framed photography as a violent act that "objectifies" those it represents. Bresheeth (2006), one of many to repeat this perspective, wrote of "a panopticon of pain and suffering, available round the clock" (p. 58). Earlier, Debord (1977) argued that we have become entirely dependent on vision to understand reality, and in *The Gulf War Did Not Take Place*, Baudrillard (1995), extending this view further, wrote that the media images of the Gulf War were unrelated to what war is really like. Against the view of a hyper-real world, Sontag (2003) asserted that to describe reality as a spectacle "universalizes the viewing habits of a small, educated population living in the rich part of the world, where news has been converted into entertainment" (p. 110).

Perhaps the role of visual images has been overstated, even in that rich part of the world. Images seem to have an impact on the individual and the public mind, but their effect is mediated by multiple and preexisting factors. Hallin (1989) and Perlmutter (1998) indicated that photographs by themselves do not influence directly. Hallin argued that public opinion had already changed before the most disturbing pictures of the Vietnam War were published, and that these images became important only when the war was almost over. Perlmutter (1998) said similarly that the influence of a photograph is largely construed after its publication. He acknowledged that pictures can have considerable effects "but only if the fire they fan is already burning—only if those effects follow pre-existing channels of prejudice" (p. 134). Any understanding of photographs as straightforward silver bullets that shape the mind of viewers, then, must be rejected.

Although Lacey (1998) concluded that the "status of the photographic images as the bearer of truth is now truly compromised" (p. 224), photography and photojournalism remain relevant. Photojournalistic styles, topics, and channels of distribution have expanded in various ways, and the number of practitioners outside Europe and North America is growing. Professional digital photographers can now compete with broadcast breaking news as the time gap between the production and consumption of images that existed with analogue photography is annihilated. With high-definition video built into more photo cameras, an increasing number of photographers will probably compete even more directly with traditional broadcast news in the near future.

The arrival of digital technology, which has increased awareness of the construed nature of photography among the public, has affected but not disintegrated photojournalism and its culture of objectivity. News agencies and publications have introduced or strengthened guidelines to protect the indexical quality of the photograph. They have received digital technology as a threat to the "objectivity" of the indexical photograph, often implicitly assuming that manipulations occur only *after* the act of photographing.

Even though new technologies are agents of change, they also are strongly dependent on the social and cultural contexts in which they are used and developed (e.g., Bijker, Hughes, & Pinch, 1987; Law & Hassard, 1999). Rather than simply replacing earlier practices, new technologies are related to old practices. For instance, in the areas of montage and collage, where digital techniques have made production much easier, older work by John Heartfield, Marinus Jacob Kjeldgaard, and their predecessors remains a point of reference. The same applies to conflict photography. Although photographers have generally covered the conflict in Afghanistan with a digital camera, they have relied on vocabularies of war photography established during major past conflicts such as the American Civil War (1861–1865), the Spanish Civil War (1936-1939), and the Vietnam War (1959-1975).

From this point of view, it is difficult to accept claims that photojournalism is dead. In the 1950s, former Museum of Modern Art curator John Szarkowski argued that "the photojournalism experiment was on its last legs" (cited in Panzer, 2005, p. 9) and others consider 1972, the year that *Life* folded, the last year of photojournalism.[10] Although the days of the illustrated magazine as a mass medium with millions of readers were over in the 1970s, new magazines and newspaper supplements continued to appear and kept photojournalism economically viable. New photo agencies were established throughout the 1960s, 1970s, and 1980s: Gamma, Sipa, Sygma, and Contact Press Images, and later Vu, Editing, and Métis. By the end of the 1980s, competition from wire services increased, and many of these agencies ran into difficulties. This eventually led in the 1990s to a consolidation in the field and the emergence of Corbis and Getty Images as global players. Around the same time, photographers countered this trend and began to use Internet technologies to set up small enterprises based on new forms of cooperation.

Commenting on the future of newspapers, Franklin (2008) observed "the most striking feature of the new compact paper editions is the *increased* use of photography for telling stories, especially on the front page above the fold" (p. 313, italics added). Print newspapers now show more and bigger photographs, and they increase the number whenever important events need to be covered (Zelizer, 2004). In a competitive market, front-cover photographs need to attract the eye of the newsstand visitor, and cover images are likely to be selected to brand and position the newspaper. Online newspapers continue to use photographs as points of attraction too, and they have introduced new ways to present them into photo-essays. An early example is Gilles Peress' "Bosnia: Uncertain Paths to Peace" (1996) on the Web site of *The New York Times*, a project that included forums for discussion and a wide range of references and resources.

Photographs regularly appear on the television news as well, and play an important role in the convergence of old and new media. They no longer are just made by the camera but by the computer and the mobile phone, from where they can be sent rapidly around the world. At the same time, traditional humanistic photography in black and white survives as a niche market, for example, in the large-scale projects of Sebastião Salgado and in work assigned by nongovernmental organizations (NGOs) such as Médicins Sans Frontières (Doctors Without Borders).

Theoretically, photojournalism need not be reduced to "pictures on a page" (Evans, 1978) or "words and pictures" (Hicks, 1952). Neither does it have to be understood in terms of a narrative sequence of images (Freund, 1980) nor as a specific medium (Szarkowski, 1966). Hariman and Lucaites (2007) defined photojournalism as a conventional form of public address that "recreates recognizable moments of social performance" (p. 32). They stressed the performative function of photojournalism, its contribution to the creation of a common life and identity. Defined in this way, as a form of popular culture, photojournalism appears to have extended its reach in the last decades.

Photojournalistic picture stories have increasingly found their way into photography museums, exhibitions, galleries, and auctions, and the number of photo-books with a journalistic angle has never been greater.

The move of photojournalists into new areas is driven to some extent by the current deterioration of working conditions and financial rewards in traditional news. For economic reasons, photographers have moved from the press to the art world, and from photojournalism to commercial photography. This is not an entirely new phenomenon. Photographers have moved between these areas at least since Edward Steichen (1879-1973), who was at various moments in his life an art photographer, a commercial photographer, and a propagandist. Today, the work of Luc Delahaye illustrates the blurring between photojournalism and art. Delahaye went to Afghanistan as a contract photographer for *Newsweek*, which resulted in a photo essay published by the magazine as "The Fall of the Taliban" (2001a). He also exhibited images made during the same trip at the Ricco/Maresca Gallery in New York, including a controversial image of a dead Taliban soldier (Delahaye, 2001b).[11] Delahaye's images were circulated through various other channels too, including museums and books such as *Arms Against Fury* (Dannin, 2002). This use of different distribution channels is not unique. Delahaye's former colleagues at Magnum Photos have mimicked their own photojournalistic style to produce advertising work, and the photo agency now sells its pictures for the promotion of commercial brands through OnRequest Images. Magnum Photos also uses books, exhibitions, and print sales to generate revenue, as do other agencies. The company that blurred the line between advertising and photojournalism most is probably Benetton. In its own magazine, *Colors*, and on its billboards, Benetton has used controversial topics such as aids, immigration, and the burqa to create its brand image.

Another example of convergence is the integration of amateur and professional photography in projects such as *Here is New York* (2002), an exhibition and book about 9/11, and *Baghdad Calling* by Geert van Kesteren (2008), who used mobile phone images and digital amateur snapshots by Iraqi citizens in an attempt to reveal backstage experiences (Goffman, 1971) to which professional journalists usually have no access. Amateur shots were used in the mainstream press to cover the 2004 Tsunami, Hurricane Katrina in 2005, and other global media events. Amateur photography as a form of citizen journalism has led to the creation of new agencies and wire services such as Demotix, but some have already closed their businesses as well.

The beginnings of the conflict in Afghanistan coincided with these and many other new developments in photography and the media. Afghanistan was the first war of the videophone, a device that allowed journalists to report without a satellite uplink (Wasserman, 2001). It also was the first war in which photojournalists had to work with digital cameras to get their images out of the country (Haviv, 2002b). And in 2001, for the first time in history, the United States succeeded in blocking commercial satellite images of a war zone when the U.S. National Imagery and Mapping Agency acquired the exclusive commercial rights to all images of Afghanistan made by Ikonos-2.

Although this study is concerned with photo reality rather than with actual events, some knowledge about Afghanistan is useful for understanding the discussion in Chapters 2 through 4.[12] To begin with, it is worthwhile to note that the word "Afghanistan" has never corresponded to a common national identity in any unproblematic way. The Afghan nation and state are the result of the competition between the British and Russian empires, and later between the Soviet Union and the United States. The term *Afghan* originally was used to designate the Pashtuns, who constitute Afghanistan's largest ethnic group. The Hazaras, Tajiks, Uzbeks, Aimaqs, Turkmens, Baluchis, and Kirgiz are other major macro-ethnic entities. But Afghanistan's population is more diverse than these labels suggest, and probably between 50 and 200 ethnic communities exist in the country (Crews & Tarzi, 2008a). Furthermore, the relation between ethnicity and identity is by no means straightforward. People may resist ethnic identification and prefer a more local form of identity formation. The Tajiks, for example, prefer to name themselves after the valleys from which they come: Panjshiri, Mujani, and Andarabi (Allan, 2003).

The homogenization of the Pashtuns started under the rule of Ahmad Shah Durrani (1722-1772), an Abdali Pashtun who founded Afghanistan in 1747. His empire included present-day Afghanistan and Pakistan, and parts of eastern Iran and western India. When the British Empire imposed the Durand border in 1893, the Pashtun population was split between Afghanistan and Pakistan.

The Durrani monarchy and Pashtun elites ruled Afghanistan until the coup of 1973 by Muhammad Daoud. He gradually reduced reliance on the Soviet Union for military and economic support. This and the fear of a rise of Islamic forces probably led to the Soviet invasion of April 1978, an intervention that was condemned internationally and displaced millions of Afghans, caused many casualties, and seriously strained the Soviet economy. The United States, China, Saudi Arabia, and Pakistan covertly supported the resistance of *mujahedin* (*holy warriors*) with weapons and money. Eventually, on April 14, 1988, Soviet President Mikhail Gorbachev agreed to withdraw from Afghanistan. The weak government of Muhammad Najibullah fell and a new *mujahedin* regime, led by the Tajik Burhanuddin Rabbani, was installed. But his regime suffered from internal divisions, and these led to civil war (S.M. Tarzi, 1993). As Kolhatkar and Ingalls (2006) wrote, if Afghanistan ever deserved to be called a "failed state," it was probably in these years, between 1992 and 1996 (p. 21). This period came to an end when the Taliban seized power.

The term *Taliban* (*religious students*) refers to its religious base but the movement has social, political, and ethnic roots, too. Many Taliban grew up as refugees in Pakistan during the Soviet time and were educated in madrasas (religious schools). The origins of these schools can be traced back to Sunni Muslim reformers of the 19th century, who founded the madrasa of Deoband in northern India after the Revolt of 1857, when Indians failed to gain independence from the British. The madrasa of Deoband gathered anti-British Muslims and proposed a return to old interpretations of Islam (Taqlid) and strict adherence to the Hanafi school of Islamic jurisprudence. The Deobandi movement's political wing, Jamiat Ulema-e Islam (Association of Islamic Scholars), was founded in 1945. The original Taliban were largely socialized and enculturated in Jamiat Ulema-e Islam madrasas, where they developed an identity based on the utopian vision of a pure Islamic state (Dorronsoro, 2005). Because the madrasas offered free food, accommodation, and education, they attracted many children of poor families, and orphans. The schools received financial support during the Cold War from the United States, Europe, Saudi Arabia, and others as part of their anti-Soviet strategy.

When they came to power in 1996, the Taliban established a patriarchal and theocratic system and reconnected with the Pashtun ethnic heritage (Crews & Tarzi, 2008a). All but one member of the Taliban's supreme *shura* (*council*) were Pashtuns (B. R. Rubin, 1999). Their rule sharply contrasted with that of the communists, who abolished tribal marriages and customs, allowed women to occupy important posts, and introduced general education for both the genders.

Like the Soviets before them, the Taliban sought control over the Afghan population, and especially women, through regulation of the physical body. Women were forced to wear the *chadri*, men had to grow their beard and wear a turban, and public beatings were introduced for those who did not follow regulations. Schools for girls were closed; women were forced out of their jobs; dancing, music, balloons, kite flying, chess, film, and photography all were forbidden. Groups such as the Revolutionary Association of the Women of Afghanistan resisted the situation in a civil way, and military resistance came from the National Islamic United Front for the Salvation of Afghanistan, popularly known as the Northern Alliance, a group of combatants of various ethnic origins supported by Iran, Turkey, France, India, Russia, and most of the Central Asian republics. The United States did not openly reject the Taliban government at first; however, this changed after the bombing of two of its embassies in Kenya and Tanzania in 1998. These led to a retaliatory missile attack, and later the UN Security Council imposed sanctions on the Taliban regime. When the attacks in New York and Washington on September 11, 2001 occurred, NATO supported the United States and invoked Article 5 of its charter, stating that terror attacks on the United States constituted an attack against all of its members. The UN Security Council demanded the hand over of bin Laden and the closure of all Al-Qaeda training camps in Afghanistan. In response, the Taliban requested evidence of bin Laden's involvement, and refused to hand him over in the absence of such evidence.

The United States dismissed attempts by Taliban moderates to reach an agreement and launched Operation Enduring Freedom on October 7, 2001. The official goal was to oust the Taliban regime from power, to destroy Al-Qaeda camps, and to capture Osama bin Laden. The campaign began as a massive attack and involved contested carpet, cluster, and daisy-cutter bombing. Several protests against the military action were organized around the world, both before and during the bombing. The first strikes were directed at Kabul and other major cities. In Kabul, Afghanistan's only radio station, Sharia Zhagh (Voice of the Islamic Law), was destroyed; in Herat and Kandahar, the airport was hit; and in Jalalabad military training camps were bombed. The military mission was accompanied by the dropping of aid packages for refugees, a contested operation. The first aid packages were hard to distinguish from unexploded cluster bombs, and the distribution of medicines for use without consultation was said to be dangerous. Médicins Sans Frontières called it propaganda (CNN, 2001) and Action Against Hunger dismissed it as a marketing stunt (Tyler, 2001).

On November 9, 2001, Northern Alliance troops took over Mazar-e Sharif. This turned out to be the beginning of the end of the Taliban regime. Tajik elements of the Northern Alliance took control of Kabul, and shortly thereafter the cities of Herat and Jalalabad fell. Al-Qaeda soldiers gathered in the Tora Bora cave complex, southeast of Jalalabad, which the United States started to bomb on November 16, 2001. On December 9, 2001, the Taliban regime came to an end when Taliban leader Mullah Muhammad Omar ordered his troops to leave Kandahar.

While the bombing campaign continued, a political reconstruction program was established in Bonn. The Bonn Agreement provided for the installation of an Interim Authority, the organization of a *loya jirga* (*grand assembly*), the appointment of a Transitional Authority, the adoption of a national constitution, and the democratic election of a National Assembly. In spring 2002, the International Security Assistance Force started as a modest operation to protect Kabul and its Interim Administration (Maloney, 2007). When the conferral of power to the Afghan transitional government followed in June 2002, the armed conflict came to its official end. A general amnesty allowed many Taliban to escape imprisonment and prosecution and to participate in the elections (A. Tarzi, 2008). Because he was the only candidate with nationwide exposure, Karzai easily won.

Various groups that claimed to be Taliban, but whose links with the original Afghanistan's Taliban Islamic Movement often remained unclear, increased their offensive against allied troops from 2004 onward. As the insurgency rose, so did the number of air strikes that compensated for the limited numbers of soldiers on the battlefield (Korski, 2008; Maloney, 2007). The civilian and military death tool grew. Amnesty International and Human Rights Watch published several reports that documented how Afghan civilians were caught up in the armed struggle, including members of Pashtun communities in northern and western Afghanistan, who became the subject of apparently retaliatory attacks. Various reports (e.g., Independent Panel on Canada's Future Role in Afghanistan, 2008; International Crisis Group, 2008; Naumann et al., 2007) expressed concern, critique, and uncertainty about the nature of the ongoing military and humanitarian mission in the country, now more frequently and less euphemistically called "war" again. In 2008, failure in Afghanistan was seen as "a realistic prospect" (Korski, 2008). Ahmed Rashid (2008) described the situation as a "descent into chaos," and *The Financial Times* concluded that "The Dream of Afghan Democracy is Dead" (Lieven, 2008). At the end of 2009, President Barack Obama announced the deployment of 30,000 more U.S. troops in Afghanistan and that troops would begin to withdraw by 2011.

## NOTES

1. The term *photo-text* is derived from Mitchell's (1994b) notion of "image-text" and borrowed from a book called *Phototextualities*, edited by Hughes and Noble (2003) and associated with "The Photo-Text conference." I use it here in a general sense to denote the intimate relationship between photographs and texts.
2. For a comparison between the liminal and the liminoid, see Turner (1974, 1982).
3. Scholarly work in the field of peace journalism includes Galtung (2000, 2002, 2006), Kempf (1996, 1997, 1999), Lynch (1998), Lynch and McGoldrick (2005), and Wolfsfeld (2001). See also J. Becker (1982), Hackett (2006), Hanitzsch (2004), Juluri (2005), S. Lee and Maslog (2005), Levin (2003), Tehranian (2002), *Conflict and Communication* Online at http://www.cco.regener-online.de, Transcend International: A Network for Peace and Development at http://www.transcend.org, and *Global Media Journal* at http://globalmedia.emu.edu.tr.

4. For Galtung (2000), truth is multiperspectival: "I'd like to see objective journalists. By that I mean journalists who are able to cover all sides of the conflict, and again I repeat: it doesn't mean you have to like them all but for heaven's sake, tell it in their own words, don't interpret" (p. 163).
5. On the domestication of news, see Clausen (2004); Cohen, Mark, Roeh, and Gurevitch (1996); Hallin and Gitlin (1993); Rolston and McLaughlin (2004); Sreberny-Mohammadi (1991); and Williams (2003).
6. The USAID photographs were collected from the organization's Web site photo gallery (72 photos organized in seven sequences), the USAID publication *Afghanistan Reborn* (36 photos), from the USAID exhibit *Out of the Shadows* (39 photos), from 69 USAID weekly reports issued in 2003 and 2004 (approximately 300 photos), and from the Telling Our Story section on the USAID Web site (40 photos). The photographs were made after the return of the USAID to Afghanistan in January 2002, and cover the period to 2005. In this period, the USAID focused on emergency relief and the introduction of basic services. The DoD images were taken from www.defendamerica.mil and www.defenselink.mil. The DoD sample included 187 photos of Operation Enduring Freedom found at Defenselink, dated between October 7, 2001 and November 29, 2005, and approximately 500 images of the "war on terror" found at Defend America, dated between October 2001 and December 2002.
7. The PSYOP images were taken from the collection of Herbert A. Friedman (2002), a retired sergeant major, who is the PSYOP history editor of the Psychological Operations Association journal *Perspectives*. The sample consisted of 51 PSYOP leaflets of an estimated 100 used during the initial stages of the intervention in Afghanistan. The texts cited from leaflets in this book are based on Friedman's. The original texts are in Dari, Pashto, or both.
8. The United Nations Development Program (2007) estimated that the adult literacy rate in Afghanistan is at 23.5% of the total population.
9. Indexicality refers to photography's referentiality and its related status of objective record. Because photographs are no longer necessarily indexical but can be entirely construed on the computer, the present era has been called "post-photographic" (von Amelunxen, Iglhaut, & Rötzer, 1996). For an introduction to indexicality, see Peirce (1958) and Lefebvre (2007).
10 The year that *Life* came to an end is probably better characterized as the end of photojournalism's modern age. Modern photojournalism began with the invention of the Leica and Ermanox cameras in the 1920s and developed into a specific form of storytelling in illustrated magazines such as the *Berliner Illustrirte Zeitung*, the *Arbeiter Illustrierte Zeitung*, and the *Münchner Illustrierte Presse* in Germany, and *Vu* in France.
11. The Afghan pictures were part of an exhibition called *History* (2003). Duganne (2007) concluded that differences between *History* and the

images in *Newsweek* depend on accompanying texts, image size, choice of camera format, and corresponding level of visual detail but not on choice of content matter, formal structuring or moral position.

12. More background information on Afghanistan can be found in Crews and Tarzi (2008b), Goodson (2001), Marsden (1998), Nojumi (2002), Rashid (2000, 2008), Roy (1995), B. Rubin (1995, 2002), and Vogelsang (2002).

# 1

# Photographs and Frames

Taking pictures involves framing, a process of selection and focus with which all photographers are familiar. This process is likely to have some impact on interpretation, although photographs are open structures that generate a multiplicity of meanings. When photographs are put to use, texts and other frames typically limit this multiplicity. This chapter discusses some of the ways in which the dual process of visual and textual framing can be understood. First, some major approaches to photography and framing are briefly introduced. Second, social semiotics and frame scholarship are contrasted to define framing more accurately in relation to culture and ideology. Third, the use of social semiotics for textual and visual frame analysis are discussed and illustrated.

## PHOTOGRAPHY AND THEORY

As far as photo theory can be mapped on a linear time line, there seems to have been a move from *what is framed* to the *formal characteristics of the photographic frame itself* and *the forces that frame the photograph*. To put it differently: Photo-theoretical developments have shifted in the course of time from preoccupations with outside reality to considerations of medium-specificity and later of institutional-discursive aspects "outside the picture." The first historical position, a strong preoccupation with outside reality, is found in the early work of William Henry Fox Talbot's *The Pencil of Nature* (1969), originally published in 1844, and in theoretical discussions by Kracauer and Benjamin. Kracauer's "Photography" (1980), first published in 1927, made a distinction between art and photography on the basis that the latter could never be free from its "outer circumstances" (p. 260), and in "A Short History of Photography," first published in 1931, Benjamin (1980) said that photography was capable of revealing aspects of reality not visible without the camera. In the 1960s, Museum of Modern Art curator John Szarkowksi shifted attention from *what* a photograph reveals to *how* it depicts reality. Szarkowski contributed significantly to the establishment of photography as modern art through a focus on the formal characteristics of the work of photographers like Lee Friedlander, Garry Winogrand, and Diane Arbus in the influential exhibition *New Documents* (1967) at the Museum of Modern Art in New York. In T*he Photographer's Eye*, Szarkowski (1966) listed five medium-specific characteristics that he believed were central to their work: "The Thing Itself," "The Detail," "The Frame," "Time," and "Vantage Point." In the 1980s, many scholars and critics began to reject medium-specificity as a form of essentialist thinking, and started to focus instead on the institutional and discursive aspects of the production of photographic meaning. This led to theories of "photographies" in the plural: photography as determined by its various uses in the ideological contexts of capitalism, popular culture, science, and the arts. Crimp (1989) described how photographs that were transferred to the Arts, Prints, and Photographs department of the New York Public Library acquired new meanings as they became objects of art; Phillips (1982) focused on the creation of photography as an art form at the Museum of Modern Art in New York; and Burgin (1982a) and Tagg (1988) discussed the role of the state in definitions of photography. Bolton collected some of these and other writings of the 1980s in *The Contest of Meaning* (1989), presenting the reader with critical histories that highlighted the discursive-constructivist aspects of photography. Reflecting the critical turn in the humanities and social sciences of the 1970s and 1980s, this project was not merely academic but was motivated by a political desire to liberate photography from the state. Photography emerged from these writings as an instrument and effect of the oppressive state apparatuses (Althusser, 1971), and as a power/knowledge tool (Foucault, 1980) that reproduced gender differences, class distinctions, and national identities. Photography was considered a tool of ideology used within "a contest of meaning" and, as Barthes (1980) suggested, an important naturalizing force.

## THE CONCEPT OF THE FRAME

In this book, the concept of the frame is applied to images and to texts surrounding the image. The notion has been used to study differences in the presentation of reality and its effects on individuals and the public.[1] In communication studies, it is closely related to the concepts of agenda setting and priming. Framing refers to the presentation of elements within messages, and the way this influences people in their understanding; agenda setting describes the selection of subject matter and its effects; and priming the prominence of evaluation criteria and its effects. Price, Tewksbury, and Powers (1997) described frame effects as *applicability* effects: Frames influence thinking by promoting the application of one concept to another. Agenda setting and priming have an *accessibility* effect: Agenda setting influences people by making certain topics more accessible in memory, and priming makes certain evaluation criteria more accessible. In short, agenda setting has to do with *what people think about*, priming with the *criteria people use to make judgments*, and framing with *how people think about an issue* (McCombs, Llamas, Esteban, & Rey, 1997).

Entman's often cited definition (1993) lists four social functions of framing: Frames are used to make issues important, to express judgment, to assign responsibility by specifying causes of problems, and to indicate actors that should or might offer solutions. Entman said that "*to frame is to select some aspects of a perceived reality and make them more salient in a communicating text, in such a way as to promote a particular problem definition, causal interpretation, moral evaluation, and/or treatment recommendation*" (p. 52). Although a single frame may not perform all these functions, Entman's definition indicates that framing is more than making some attributes of an object salient, a process also known as second-level agenda setting (McCombs et al., 1997). The attribution of causal and treatment responsibility appears to be a particularly important aspect of framing. These aspects have been called diagnostic and prognostic framing (Benford & Snow, 1988; Snow & Benford, 1992). Diagnostic frames provide causal explanations, whereas prognostic frames suggest remedies.

In their famous experiment, cognitive psychologists Kahneman and Tversky (1984) demonstrated that a slight shift in framing a remedy can radically alter the preferences for it. Similar effects have been demonstrated in other areas and in nonexperimental settings (Iyengar, 1991). Iyengar found that the way people, events, and circumstances are presented has a significant effect on conclusions about responsibility. He suggested that television journalists frame the news more often in *episodic* terms, with an emphasis on individuals and events, rather than in *thematic* ways that discuss trends, contexts, and issues in general or abstract terms. As a result, news audiences are inclined to attribute responsibility to individuals rather than societal factors, and this determines their preference for solutions to a problem. *Episodic* frames typically suggest that solutions are located at the level of the individual person, whereas *thematic* frames invite the public to think about solutions that focus on the contextual conditions of a problem. Because *episodic* framing does not encourage people to think of solutions at the societal level, it is more likely to support the social status quo and the existing establishment. Lakoff (2006) related these differences between direct causation by an individual agent, on the one hand, and systematic causation within a wider system, on the other hand, to differences between conservative and progressive politics in the United States.

Since Kahneman and Tversky (1984), many frame scholars have accepted a relatively strong effects perspective. Entman (1993), for example, concluded: "If the text frame emphasizes in a variety of mutually reinforcing ways that the glass is half full, the evidence of social science suggests that relatively few in the audience will conclude it is half empty" (p. 56). But text frames are not independent determinants. Frame effects are mediated by political and cultural knowledge (e.g., Domke, Perlmutter, & Spratt, 2002; Zaller, 1996). This posits framing as an effects theory between models of media hegemony and models of audience autonomy. It is assumed that people are most likely to accept message frames *when they fit their beliefs and values*. The level of fit between message frames and audience frames, known as "frame resonance" in social movement literature (Benford & Snow, 1988), determines whether a message frame is accepted, negotiated, or rejected.

*Generic* or *structural* frames that control a broad range of news messages are the already mentioned *episodic* and *thematic* frames (Iyengar, 1991), and the *conflict*, *human interest*, and *consequences* frames (Price et al., 1997). The latter three suggest that journalists prefer to represent reality in terms of oppositions, that they tend to focus on people and events to make issues less abstract, and that they often will try to make events more relevant by indicating their impact. Elsewhere, this range of *generic* frames was extended to include *attribution of responsibility* and *morality* (Semetko & Valkenburg, 2000). Others identified the *ethno-nationalist* frame, the *liberal-individualist citizenship* frame, and the *harmony with nature* frame as "*master frames* that enjoy high cultural resonances" (Koening, 2006, p. 64). Beyond these *generic* frames, various researchers identified more specific frames that are applicable to a specific event or a set of events. The *Cold War* frame can be understood as a more specific version of the *conflict* frame, and the *falling domino* frame as a more specific version of the *Cold War* frame and the *consequences* frame (see also Wolfsfeld, 1997). Frames are definitions of reality that differ from each other not only in their specificity but also in their degree of clarity, complementarity, permeability, durability, frequency, and spatiotemporal spread. This makes it sometimes difficult to distinguish analytically between frames, although many frames co-occur with clear counterframes. Because a frame indicates what aspects of reality are deemed important or unimportant, and suggests how reality should be interpreted, competing frames represent a "contest of meaning," an ideological struggle over the meaning of certain issues, events, and people. When Tuchman (1978) and Gitlin (1980) introduced framing in communication studies, they highlighted this ideological aspect. Both saw framing as an unconscious hegemonic mechanism. The hegemonic model, based on the concept introduced by Gramsci (1971), suggests that the great masses give their consent to the elites spontaneously. The media are seen as "ideological apparatuses" (Althusser, 1971) that rely on government and elite sources routinely and thus provide the public primarily with the dominant frames of the politically powerful. The term *hegemonic* suggests that they perform their work in an unobtrusive and unconscious way, whereas *ideology* usually denotes a more explicit and deliberately activated system. Media researchers adhering to the hegemony perspective believe that the news is a naturalizing force that produces consensus by presenting the dominant view as the normal one, and generally think of audience members as mostly passive recipients of news. The manufacturing consent or propaganda model (Herman & Chomsky, 1988) proposes a version of hegemony in which capitalism and economic power are identified as the source of media and government frames. The indexing model (Bennett, 1990), another variant of the hegemony model, holds that "mass media news professionals, from the boardroom to the beat, tend to 'index' the range of voices and viewpoints in both news and editorials according to the range of views expressed in mainstream government debate about a given topic" (p. 106). According to this view, the media may present the public with various frames but this will only happen when the political elite is in disagreement over an issue and in a highly selective and ultimately even merely ritualistic way. For Entman (2004), the hegemony and indexing models are "based largely on events during the Cold War" (p. 4) and reflect the bipolar perspective of that era. Entman's own cascade model is presented as a more complex alternative view on the relation between politics, the media, and the public. According to Entman, news frames cascade down from the White House to political elites, to the media, and then to the public. Whereas the hegemony and indexing models hold that the political field determines the news, the cascade model suggests that politicians also are constrained by the media, and that public opinion guides the media. The hierarchy of influences model, proposed by Shoemaker and Reese (1996), adds individual agency to this perspective as an important mediating factor. Bennett and Livingston (2003) further noted that events with a great immediacy increasingly determine the news too, although they also said that official sources remain "as much a part of the news as ever" (p. 376). In short, frames and their effects are determined by multiple factors.

## USING SOCIAL SEMIOTICS FOR FRAME ANALYSIS

Although some researchers may rely solely on keywords to identify media frames, most will use a larger set of framing devices for their analytical purposes. Methods based on the mere counting of keywords or metaphors, without contextual disambiguation, presuppose a simple one-to-one correspondence between signifier and signified. A more elaborate analytical technique is required to capture meaning in context.

Pan and Kosicki (1993) considered syntactical, script, thematic, and rhetorical structures as framing devices, and Gamson and Modigliani (1989) made a distinction between *framing devices* that regulate the presentation of an issue (metaphors, exemplars, catchphrases, depictions, and visual images) and *reasoning devices* (causes, consequences, and moral claims) that explain an issue (see also Gamson & Lasch, 1983). My approach is similar but grounded in social semiotics. There are several reasons why I think that social semiotics is suitable for media frame analysis. First, it shares with frame research in social movement studies the theoretical idea that communication is based on strategic resources (i.e., *framing devices*) that groups in society deploy to get their messages across and to exercise influence. Second, it offers the frame scholar a coherent toolbox that is suited for the analysis of texts *and* images. Third, the toolbox provides the researcher with a systemic overview of the resources that individuals can choose from to build frames. Fourth, both semiotics and frame studies offer a double perspective on meaning: a relational and a referential one. It is well known that de Saussure (1983) considered the sign predominantly from the relational perspective—that is, from within the system of signs—and that Peirce (1958) integrated the referential world in his theory of semiosis. Frame studies and social semiotics refer to both aspects. Snow and Benford (1992), for example, included the referential world in their definition of the frame as "an interpretive schemata that signifies and condenses the '*world out there*' by selectively punctuating and encoding objects, situations, events, experiences, and sequences of action in one's present or past environment" (p. 137, italics added). Gamson and Modigliani (1989) referenced the outside world as well but seemed to stress the relational aspect:

> media discourse can be conceived of as a set of interpretive packages that give meaning to an issue. A package has an internal structure. At its core is a central organizing idea, or *frame*, for making sense of relevant events, suggesting what is at issue. (p. 3)

Frames are defined here as conceptual nodes within a wider network of related meanings, which Gamson and Modigliani called "interpretive packages" and which classic semioticians would call "codes." Elsewhere in frame studies too, the focus is on the "structure in which meaning is contained in and conveyed by the relationships among the elements" (Williams & Benford, cited in Benford, 1997, p. 413). As in Saussurian semiotics, it is assumed that signs gain their meaning from their position in a larger system. Social semiotics replaced this focus of classic Saussurian semiotics on signs within the language system with an emphasis on the deployment of semiotic resources in social life, and this closely resembles the approach to framing taken by scholars in social movement research (e.g., Benford & Snow, 2000). Understanding framing devices as resources shifts the emphasis from the unconscious production and reception of frames to their (partly) conscious application. This shifts seems justified given the presently large number of professional frame sponsors such as spin doctors, public affairs professionals, and PSYOP specialists.[2]

The deployment of resources is related to Peirce's notion of the *interpretant*, the semiotic concept that most closely resembles the notion of the frame. The interpretant is an idea for understanding a *representation*. In the definition by Peirce, a sign consists of a representamen (de Saussure's *signifiant*), an *interpretant* (de Saussure's *signifié*), and the referent (object) to which both representamen and interpretant refer. Interpretants are representamen too, which explains the existence of "floating chains of signifiers." Social semioticians accept that signifiers can float, theoretically ad infinitum, but that this is pragmatically limited by social context. Most importantly, for Peirce and social semioticians, meaning is a dynamic process: An interpreter has to choose an interpretant (*mental frame*), to make sense of a situation (*the world out there*) that is being represented (*media frame*).

Historically, the origins of social semiotics can be traced back to the work of the Polish anthropologist Malinowski (2002), who said that language cannot be studied without an understanding of its cultural context. Firth (1957) took this idea further and argued that contexts predict language use, and vice versa. Halliday (e.g., 1966, 1978, 2004) presented a systematic account of this language-context correlation in systematic functional linguistics. He identified three major contextual determinants of language use: ideology, genre, and register. A consideration of framing from the point of view of these three determinants helps to define framing and its relation to culture more clearly.

Referring to culture as "deeper layers of communication and thinking" (Reese, 2001) or a "broader worldview" (Gamson, 1988), frame scholars seem to think of culture as a given and of framing as something that essentially occurs *within the human mind.* With the exception of social movement frame research, culture in frame studies usually appears as a more or less static repository of beliefs and ideas. Social semiotics understands culture differently and more dynamically, as does social movement frame research. Social semiotics associates culture with genre, which it defines as habitualized patterns that people rely on to achieve a certain communicative purpose. Genre is a form of *social interaction* that interfaces between ideologies (i.e., value systems) and registers. Social semiotics understands genre as a realization of ideology, and register as a realization of genre. Bottom–up frame analysis typically starts with identifying the actually deployed framing devices at the register level and then moves up to the more abstract levels of genre and ideology. The realizational relationship can also be studied in a top–down way to verify whether known abstract values are realized at the lower levels.

## THE PHOTOJOURNALISTIC REGISTER AND GENRE

Register refers to three variables of the situational context of a communicative act: field, tenor, and mode. First, *field* is central in definitions of framing that stress the referential and structural nature of framing. It is the knowledge that people establish about the "world out there" through specific configurations of processes, participants, and circumstances, and by imposing logical relations of time, space, and cause unto these. Given the syntactical implicitness of photography, logical relations within a photograph typically remain open and ambiguous, whereas the experiential components are concrete and analogous to reality. The three major constituent categories of field suggest a basic threefold typology of ways of framing the world: *process-oriented*, *participant-oriented*, and *circumstance-oriented. Process-oriented* and *participant-oriented* frames are roughly congruent with the *human-interest frame* (Price et al., 1997) and Iyengar's (1991) concept of episodic *framing.*[3]

Second, *tenor* refers to the relational aspect of a communicative act. This dimension was of central importance in Goffman's *Frame Analysis* (1974) and Bateson's "A Theory of Play and Fantasy" (1972), two foundational texts for frame studies. Bateson's famous example is that of two monkeys fighting in the Fleischaker Zoo in San Francisco. According to Bateson, the monkeys communicated the meta-message "play/not for real" by pulling their punches and thus defined a specific relation between them. The relational perspective was further developed in family therapy, a field that has used the concept of the frame to help patients discover new meanings and behavioral patterns (Haley, 1963). Photographic framing, seen from this perspective, is not about the referential world but about how photography structures relations between photographer/viewer and represented participants. This interpersonal aspect is grounded in *photographic behavior* a term used by Milgram (cited in Newton, 2001, p. 35) to indicate that photographers and subjects enter into a particular relationship and establish a separate plane of reality during the photographic act. According to Bourdieu (1990), all people seem to alter their behavior when they are aware of the presence of a camera.

Photographic behavior finds its way into photographs and determines the position of viewers in three major ways: type of camera angle, type of shot, and type of gaze. *Camera angles* regulate social power and involvement, with high vertical angles typically casting the viewer in a powerful role above the depicted participants and low angles realizing the reverse; horizontal angles regulate the degree of involvement, with frontality evoking engagement and oblique angles suggesting detachment (Kress & van Leeuwen, 2003). The *type of camera shot* controls the psychological distance between viewer and represented participants. Greater physical distance typically suggests greater psychological distance, and when physical distance decreases psychological distance decreases as well.[4] The *gaze* of represented participants establishes a particular type of contact with the viewer. Kress and van Leeuwen (2003) made a distinction between offer and demand pictures. Many photojournalists prefer to shoot offer pictures that position the viewer as the subject of the gaze. These images suggest a preference for the unobtrusive view and depict people gazing elsewhere. Alternatively, to enhance emotional impact, photographers may choose to create demand pictures with depicted participants looking directly at the camera. Demand pictures of victims have frequently been used to critique war. The victims seem to demand something from the viewer, and seem to invite the reader into a relationship of which the nature is indicated by a facial expression, a bodily posture or some other means. According to Panofsky (cited in Kress & van Leeuwen, 2003), demand pictures originated in self-portraits around the time when artists became self-conscious individuals. The viewer is turned into the object of the gaze of the depicted.

Finally, *mode* refers to the ways in which field and tenor are integrated. Although photographs are syntactically implicit, they must have coherence in order to be meaningful. As other semiotic systems, photography achieves this coherence through a variety of compositional mechanisms. These mechanisms coordinate representational and interpersonal meanings to create a sense of shared experience. They are important aspects of the frame as "an organizing principle."[5] A key compositional mechanism that is generally recognized as a major framing device in the frame literature is *salience*. The salience of elements within a picture depends on a number of visual factors such as relative size, lighting, position, sharpness of focus, tonal scale, color contrast, and saturation. Kress and van Leeuwen (2003) identified three other organizational mechanisms with an impact on visual meaning, and with important cultural connotations: information value, modality, and connection (which they called *framing*).

First, *information value* is determined by the placement of elements within a picture. Center and margin indicate what is more important (the center) and what is less important (the margin). Top positions typically represent generalized and idealized information, whereas bottom positions are normally reserved for specific, practical, or real information. Left positions normally contain the known, and right positions contain the new. Kress and van Leeuwen (2003) hypothesized that center-margin compositions are more common in Asian cultures with a greater emphasis on hierarchy, harmony, and continuity, whereas left–right and top–bottom compositions are more typically Western. In Western cultures based on the Roman script, known and familiar information is normally presented on the left side, and new information that requires special attention is generally presented on the right side. This traditional left–right organization in the West implies that an obliquely depicted person looking from left to right appears to be looking into the future ("the new"); the same person looking from right to left appears to be looking into the past ("the known").

Second, *modality* refers to the capability of a semiotic system to represent contents as true or possible. In Western cultures, photography typically has been associated with the truth, a high level of modality that derives from its naturalistic qualities. Kress and van Leeuwen (2003) listed eight different modality markers: color saturation, color differentiation, color modulation, background articulation, pictorial detail, depth, illumination, and contrast and brightness. These markers can be seen as scales that have a point at which the photographic representation is at its most veritable. This point seems to depend on the context in which the modality markers are used. In the photography of conflict, veracity has been enhanced by techniques such as poor focus and excessive grain that are rather unlike natural perception. Robert Capa's photography of D-Day, which earned him the title of "The Greatest War-Photographer in the World" (*Picture Post*, December 1938), is a well-known example. Grain and poor focus enhanced their status of authentic documents and reinforced their high modality. These markers seem to suggest a bodily experience beyond visual perception. In *The Skin of Film* (2000), Laura Marks argued that certain films appeal not only to seeing but to the proximal senses by highlighting the materiality of film through scratching emulsion, over- and underexposure, alternation between film and video, and the use of toy cameras. This seems applicable to Capa's D-Day photographs and to other more recent images as well. Pixelation and underexposure in Tim Hetherington's World Press Award-winning image of an exhausted American soldier in Korengal Valley in Afghanistan, for example, suggest a similar experience of bodily being in the war.[6]

Finally, the *connection* of elements within a picture may create a sense of belonging, and the *disconnection* of elements a sense of separation. Connection and disconnection can be achieved through a variety of means including dividing lines within the picture, empty spaces, color contrast, and a picture's actual borders.

The register variables of mode, tenor, and field are determined by genre. Central to the Western photojournalistic genre is the view of the experiential world as one of doing and happening rather than being. Although there exists variation across such photojournalistic subcategories as spot news and combat photography, Western photojournalists usually are taught to make their photographs dynamic by depicting action. A classic photojournalism textbook like Kenneth Kobré's *Photojournalism: The Professional's Approach* (2008) is focused on how to capture the action, and advises to shoot people in action. It encourages photographers to think of news as narrative and dramatic, and consequently to produce images that convey or suggest a causally related sequence of events. This view appears to be more Western than Eastern. The Tao principle of *wu-wei* (*non-doing*), for example, stands in sharp contrast to the Western values of achievement, progress, and productivity, and the associated focus on change and linear time.

In addition to its focus on action, the photojournalistic genre is characterized by an interest in individual human participants. This *human-interest* frame reflects and constitutes an individualist ethos, and, when combined with a focus on leadership, creates a view of the world in which responsibility is relegated to a small number of people, often elite leaders (Bennett, 1995). This emphasis on individual leadership obscures the economic, cultural, and social causes of conflict, and this may obstruct conflict resolution. Politicians and military leaders contribute to this effect when they stage "photo opportunities" (i.e., pseudo-events with themselves as protagonists).

Circumstance-oriented photographs that draw attention to the means by which actions are performed, the accompaniments of participants, or their settings are generally less preferred in photojournalism because they do not possess the immediacy and drama with which news media usually prefer to attract audiences. Artist exhibitions and documentary photo-books like Norfolk's *Afghanistan: Chronotopia* (2002) and Seawright's *Hidden* (2003) focus on the settings of war, whereas a photojournalist's photo-book like Haviv's *Afghanistan: The Road to Kabul* (2002a) is typically more action oriented. The relative lack of interest in the circumstances of events or people can be interpreted as an illustration of the tendency in Western philosophies to focus analytically on objects, people, and actions rather than holistically on the environment in which these occur (Ji, Peng, & Nisbett, 2000).

## VISUAL AND TEXTUAL FRAMES

The social semiotic analytical apparatus can be applied to images and texts. The text–image combination has been called "the most ubiquitous means of communication" (Martinec & Salway, 2005, p. 337). Their co-existence is found in the newspaper, the photographic gallery, the online photo-essay, and other forms of multimodal cultural expression. Text and discourse have been particularly important for photojournalism because photographs have traditionally become photojournalistic by surrounding them with text (Evans, 1978; Hicks, 1952). When photographs were first used in the press, as the basis of wood engravings, photographers were told to caption their images so that publishers could make sense of them. "We shall be much obliged to our photographic friends if they will write in pencil the name and description on the back of each picture, together with their name and address," wrote publisher Frank Leslie in 1860 (cited in Carlebach, 1992, p. 65). It is as if "all photographs wait to be explained or falsified by their captions," said Sontag (2003, p. 10). Benjamin (1968) was aware too that contextual frames could reorientate the meaning of a photograph, and he saw them as devices to control photography's tendency to aestheticize reality. Barthes (1977) used the term *anchorage* to describe the process through which texts provide viewers with specific cultural and moral connotations, and his idea of anchorage was later echoed by Hall (1981): "It is a very common practice for the captions to news photographs to tell us, in words, exactly how the subject's expression *ought to be read*" (p. 229). Captions are seen here as disciplinary devices that may graft ideological ideas unto the locations, people, and events depicted by a photograph.

The term *photo-text* stresses the mutual and close relationship between the photographic and the linguistic. The notion suggests that there are basically two ways to approach photographic framing: from the inside and from the outside. The first looks at the photograph as a scene and as a metonymic device showing us a slice of life. It considers the photograph as a context, unframed by the text. The second considers the contextualization of the photograph by the text that surrounds it. To combine both perspectives is to consider photo and text as one mixed media unit. This mirrors common practice: People look at images and seek additional information in talk and text to know who is in the picture, where the picture was taken, and so on.

Martinec and Salway (2005) introduced a system to describe the relationships between texts and images based on social semiotics and the work of Barthes. One relationship that is frequently found in journalism is what they called "exemplification" (p. 350). This occurs when a text generalizes a picture, or vice versa. When captions generalize depicted participants, the viewer is invited to engage with social categories rather than individuals. This tends to promote stereotyping based on categories such as gender, class, and nationality, especially when specific visual elements are routinely singled out to homogenize a social group. Categorizations may divide the world schematically into individuals responsible for what goes wrong (villains) and individuals who solve the problems (heroes). Such a division often involves a third category, the victims, who need to be saved by the heroes from the villains. In accordance with this basic schema, the original official plot line of the "war on terror" was that the innocent Afghan and Western civilians (victims) had to be saved by the Western soldiers (heroes) from the Taliban and Al-Qaeda (villains). Even though the Obama administration dropped the "war on terror" concept, it generally seems to continue to adhere to this basic scheme.

Perlmutter (1998) demonstrated how captions create a frame for understanding a photograph. In one experiment, he gave the iconic picture by Eddie Adams of an execution on the streets of Saigon to half of the students in a class of 40 with the minimal caption "Street execution, Saigon, 1968." The other half received the same picture, but with a full quote from Loan, the general in the picture: "Many Americans have been killed these last few days and many of my best Vietnamese friends. Now do you understand? Buddha will understand." Whereas the first group concentrated on the Viet Cong prisoner about to be shot and the impassivity of the shooter, the second group interpreted the picture in the light of the caption and discussed the justifications for the shooting.

A minimal caption leaves ample room for interpretation and has therefore been used widely to position photographs as forms of art rather than forms of photojournalism. Discussing caption changes across three books by photographer Don McCullin, Scott (1999) noted that captions with minimal information are "closely connected with the ageing of the photograph, its gradual accession to the status of icon, the broadening and de-historicizing of its perspectives" and that it is only when the photograph moves from being photojournalistic to documentary that it "assumes an autonomy, speaking for itself, without the prejudicial intervention of a commentator" (p. 86). Caption changes may signal political changes. The shift from "Marines tormenting Vietnamese civilian captured during the battle of Hué, 1968" in McCullin's *Hearts of Darkness* to "A Vietnamese civilian tormented by his 'liberators,' American marines" in his autobiography *Unreasonable*

*Behaviour* indicates a drift in sympathy toward the Vietnamese, according to Scott (p. 90).

That viewers rely so much on texts for the interpretation of images is related to the nature of language and the nature of the photographic system. Like language users, photographers create patterns of meaning through paradigmatic choice and syntagmatic combination (de Saussure, 1983). Available paradigmatic options are described in most photographic manuals. The options include—with further categorical subdivisions—choice of light, color, focal length, focus, aperture, speed, tonal scale, image definition, composition, perspective, subject matter, and so on. Photographers need to make mutually exclusive choices from these categorical systems: either a slow or a fast shutter speed, for example. The photograph as a syntagma is a combination of these selections, for example, a combination of small focal length, available light, color film, small aperture, and high speed. The combination of the paradigmatic and syntagmatic axes, governed by cultural rules and conventions, establishes the photograph as a meaningful layered whole of time, space, and human relations. Language is used in a similar way but one major difference is that linguistic syntagmas are arranged sequentially and processed progressively through time whereas photographic syntagmas are organized spatially and processed simultaneously. The space of a photograph cannot be divided into meaningful units the way a text can be divided in various parts, and perception of the whole photograph normally precedes the perception of its parts. Photographs have therefore been described as syntactically implicit (Messaris, 1997). As Messaris explained in *Visual Literacy* (1994), the explicit syntax of natural language allows users to formulate hypotheses, conditions, negations, affirmations, exceptions, qualifications, and various abstractions needed for complex analysis; and it also allows for the representation of all kinds of sensory experiences. Due to its analogical nature, photography can only imply some of these at best. This is why photographs depend so much on texts or other contexts for making meaning and propositions, even though images can convey a number of abstract qualities, such as contrast and rhythm, and can imply causality and linearity through juxtaposition and montage. The lack of an explicit photography syntax makes viewers less aware of the claims that images can nevertheless suggest or support through their structures and formal characteristics (Messaris & Abraham, 2001).

## TWO EXAMPLES

The two examples here illustrate how a bottom–up multilevel frame analysis of visual and textual elements can be conducted. I start from an observation of the scene (*field* in systemic functional linguistics) and then move toward the more abstract ideological level. In their fully fledged form, scenes consist of a process (action, event, state of being), a participant (human or nonhuman figure), and a circumstance (secondary participants: setting, means, and accompaniments). A photograph may display a primary scene that contains a secondary scene, which in turn may contain another subscene, and so on. Every scene gives an answer to the question of selection: *What and who is framed*? For example, "*these* wounded children in *that* hospital" is a concrete scene that consists of two functional categories (children and hospital) and one attribute (wounded). The scene could be embedded in another scene such as that of a medical examination of the wounded children.

A scene is generalized into a theme by combining an understanding of the ideational categories of the scene with an analysis of the interpersonal and compositional meanings of the photograph and its accompanying caption, and of the genre aspects realized by the photograph. A thematic meaning no longer directly refers to concrete events or people but designates a prototypical representation. Themes provide an answer to the question: *What kind of cultural prototype is applied to the depicted processes, participants, and/or circumstances*? In the example, the wounded children might be associated with the prototype of the innocent child affected by violence. Innocence and victimhood are not inherent characteristics of the depicted children but culturally specific constructions with a complex history including, for example, representations of Christ as a child and events such as the abolition of child labor. They can be implied in a photograph through such attributes as high key lighting, a high vertical angle, or a facial expression. An accompanying caption can also indicate the preferred thematic meaning. It might literally include the words "innocent victim."

Themes are further generalized into *master* frames or ideological orientations that exist across different photographs and that reveal a more foundational idea. Introduced by Snow and Benford (1992) to describe structures of meaning across a number of social movements, I apply the term *master* frame as a meta-label for paradigmatic sets that specify a unity of culturally resonant and politically relevant contents and themes. *Master* frames are defined as core ideas through constant comparison of scenes and themes. They are applied across a wide range of different scenes and themes. For example, purity is a master idea related to the theme of the innocent child victim. Master ideas serve as a general guide and principle for understanding the world. They may exist across cultures but their content will vary. For example, purity in the West is tied to the Judeo-Christian tradition and its specific understandings of body, sexuality, and social relations. Because cultures are not homogeneous entities, contradictory *master frames* across a range of representations are to be expected.

## EXAMPLE 1: THE SPIRIT OF BERLIN

The first example is "The Spirit of Berlin," a photograph from the front page of *Die Welt*, published on October 9, 2001 (Fig. 1.1). It depicts three men leaving an aircraft. The men are clad in similar clothing. The photograph is a tight shot taken with a long focal length. The focus is on the three men and the name of the airplane, "The Spirit of Berlin." The lettering is in the upper part, and the soldiers are in the lower part. The caption is "The crew of the military transport machine 'The Spirit of Berlin' has safely returned to the base Ramstein in South-Germany after dropping aid packets over Afghanistan."[7] It contextualizes the moment captured in the photograph and extends the image by adding a new, secondary narrative process, the dropping of air packets. It calls the airplane "military" but qualifies it further by reference to its humanitarian mission. A reasonable summary of the photographic scene would be *soldiers arriving by aircraft*. The caption adds temporal information to this by including *after a humanitarian mission*.

**Fig. 1.1. The Spirit of Berlin.** © Thomas Kienzle/A.P./Reporters (2001). Published on the front page of *Die Welt*, October 9, 2001

The meaning of the photograph is strongly determined by its vertical compositional arrangement. The airplane's name positioned at the top figures prominently in the photograph as a headline within the picture, and it qualifies the nature of the humanitarian mission in a dominant way. According to Kress and van Leeuwen (2003), the upper part of a vertical axis in Western culture represents the Ideal; what is placed beneath represents the Real. Following this logic, the men leaving the airplane render the ideal of "The Spirit of Berlin" concrete. We may reasonably assume that at least a considerable portion of *Die Welt*'s German audience is likely to link "The Spirit of Berlin" to the Berlin Airlift (1948-1949) when the United States and other allied forces flew cargo to Berlin to counter the city's blockade by the Soviets. Some German audience members might remember that U.S. President Bill Clinton named the aircraft "The Spirit of Berlin" in 1998 to commemorate the 50th anniversary of the Berlin Airlift.

The Berlin Airlift is known to have changed the view of the German population toward the United States. Instead of an occupying force, the United States became a liberator, savior, ally, and friend. The photograph of the men leaving "The Spirit of Berlin" proposes a reading of the intervention in Afghanistan as a form of liberation and international friendship analogous to this common interpretation of the Berlin Airlift. The historical exemplar of the airlift associates the humanitarian operation in Afghanistan with the Cold War and the immediate aftermath of World War II. The photo-text suggests that Germany under the Nazis and communists can be compared with Afghanistan under the Taliban, and that the historical role of the United States is similar in both cases. It implies that in the post-Cold War era the United States remains a guarantor of Western freedom and democracy. The use of the Berlin Airlift as a historical exemplar makes the invasion of Afghanistan part of the continuous history of Pax Americana, and of the involvement of the United States in democratic nation-building. This theme of *freedom under Pax Americana* can be related at a more abstract level to a more general belief in the positive humanitarian outcome of conflict. The photo-text encourages viewers to accept the idea that the intervention in Afghanistan will produce freedom for the Afghans in the way that World War II produced freedom for the Germans. It is congruent with the official discourse of the "war on terror," which President Bush partly defined as a just war for freedom, an idea that echoed Gen. Eisenhower's description of World War II as "a crusade for freedom."

The picture relates two historical events and suggests that history is a chain of conflicts in which the United States operates as a liberator and nation builder. The relationship between the two events, however, is mostly rhetorical since the liberation and rebuilding of Afghanistan and that of Germany hardly share any common ground. The nation-building effort in Germany involved 1.6 million troops, whereas the effort in Afghanistan has been described as "nation-building lite" (Ignatieff, 2002) because, compared with other international operations, it has involved relatively few forces. The total number was expected to rise to a peak of around 150,000 troops by summer 2010. Simple historical associations remove such specifics, and this affects the meaning of both events. The intervention in Afghanistan is not only seen through the lens of the Berlin Airlift but the Berlin Airlift and the Cold War are also reframed in the light of the intervention in Afghanistan. The Cold War is reinterpreted as another Operation Enduring Freedom. It becomes part of a history of liberation, and its violence and oppression are obscured.

In other contexts, the Berlin Airlift may well be used to create an entirely different frame. In a commentary on U.S. operations in Afghanistan, Gen. Tommy R. Franks, commander in chief of U.S. Central Command, used the Berlin Airlift as a framing device to stress technological superiority and to create a history of warfare as a chain of technological developments: "With all due respect to those who accomplished something incredible during the Berlin Airlift, it seems to me that the duration, size, intensity, and the result of those aviation efforts into and out of Afghanistan are beyond what we may have ever planned for" (cited in Grier, 2002).

In most cases, instances of historical referencing are problematic because they tend to obstruct the proper exploration of the specific structural origins of a problem and its possible solutions. But it is a popular practice in journalism. World Wars I and II have frequently been used to legitimate the armed intervention in Afghanistan and to frame it as a just war. One study found 460 Afghanistan-related articles with a reference to World War II in a single month in the *Times*, the *Telegraph*, the *Guardian*, and the *Independent* (Lewis, 2002). The *Times* published photographs of former Prime Minister Margaret Thatcher and Winston Churchill during the initial stages of the intervention in Afghanistan, associating the intervention with British successes in the Falklands/Malvinas War and World War II, and enhancing a sense of history as a chronology of leadership and war. World War II was evoked more indirectly too, for example, in photographs of the "liberation" of Kabul that were reminiscent of allied troops welcomed in the streets of European cities. British media also contextualized the intervention by turning to early 19th-century British travel writings and histories, especially in the first weeks of the intervention, when journalists had no access to the conflict zone (Fowler, 2007). Fowler qualified references to the popular writings of Rudyard Kipling in news reports at that time as "extensive and widespread" (p. 48) and as "almost standard in tabloids such as the *Daily Mail* and the *Mirror* as well as in the *Independent*, the *Scotsman*, the *Times* and the *Sunday* Times" (p. 49).

As early as 2001, various media expressed doubts about the intervention by drawing parallels between Afghanistan and the Vietnam War (e.g., Apple, 2001; Buruma, 2001), a war that generally has been used to stress the risks and dangers of a conflict. In February 2009, for example, *Newsweek* called the conflict in Afghanistan "Obama's Vietnam" (Barry & Thomas, 2009), echoing Cold War descriptions of Afghanistan as the "Soviet Union's Vietnam" (Gregory, 2004, p. 35) and colonial images of Afghanistan as an unconquerable territory.

The use of historical references for rhetorical purposes is not limited to the press. When the Museum of London ran the photography exhibition *After September 11: Images from Ground Zero* (2002-2003), which the United States sponsored, it simultaneously put images of the London Blitz, the destruction of the British capital by Nazi Germany, on show, and thus established a relationship between the two events, and between the nations involved in them. The exhibitions created a common bond and framed both nations as resilient (Kennedy, 2003).

## EXAMPLE 2: THE KABUL-KANDAHAR ROAD

The second example is a photograph from the USAID that depicts a group of Afghan men near the Kabul-Kandahar road while it is being paved (Fig. 1.2). The photograph was originally part of a short photo-essay of seven pictures. Some workers, machines, and a few small

**Fig. 1.2. "Local residents observing the paving of the Kabul-Kandahar road". USAID (2003)**

buildings, probably set up to house construction workers, are in the background of the picture. Two of the Afghan men are looking at the road and the machine in the background, whereas the others seem to be looking at the camera and photographer. There are several ideational categories that can be discerned on the basis of their functional role: the Afghan men (the participants), the road (the setting, and, for two of the men, the object of vision), and "looking" (the process). The workmen and the machine in the background are part of a secondary narrative process, the paving of the road. This secondary process is the object or goal of the first process of looking—at least for two of the Afghan men—and the primary process in the other images of the photo-essay. The two processes, observing and paving, are separated from each other by a strong diagonal line in the middle of the picture, created by the left side of the road and by the red and white railing. The contrast between the light patch of sand on the left and the dark highway on the right further suggests two separate, disconnected worlds.

Both spaces are endowed with oppositional qualities. The space on the left side evokes a traditional, premodern, local, and rural world. The men in this part of the picture wear traditional clothes, one of them uses a bicycle, and they stand on an uncultivated patch of land. The space on the right side with the highway suggests a modern and urban world. It evokes a world of speed, machinery and enterprise. The caption evokes the same idea of two different worlds: one premodern, the other modern. It mirrors the compositional splitting of the photograph in a left and right frame: "Local Afghans observing the paving of the Kabul-Kandahar road." Lexical choice contributes to the effect: The Afghan men are identified as "local," and the caption writer interprets their activity as "observing," which is "looking in order to learn." The act of observation defines them as nonparticipants in their own "developmental" process. They are portrayed as bystanders and onlookers in their own world. In the text directly below the caption, their modern future is defined as dependent on the financial support of the United States while the frequent use of nominalization ("reconstruction," "rebuilding," "movement," "development") obscures concrete human agency and stresses an institutional, impersonal approach toward development. The text suggests that the road will lead Afghanistan to civilization and progress in the economic and commercial sense of Adam Smith's liberal ideas:

> Reconstruction of the country's principal road system is the key to Afghanistan's economic recovery. Rebuilding the transportation infrastructure will enable the movement of people, aid resources, and farm and trade goods—all essential to Afghanistan's development. The United States has provided about $190 million, through USAID, to reconstruct a 300-mile segment of Highway 1. (United States Agency for International Development, 2003)

Again, three levels of framing can be discerned. The first is the photographic scene, which is summarized by the caption as *Afghan men are observing the paving of the Kabul-Kandahar road.* The second is the theme. The image in Fig. 1.2 seems to represent a prototypical understanding of development as "the white man's burden," originally Rudyard Kipling's 1898 call on the United States to support the imperial ambitions of Europe. At a deeper level of analysis, the USAID picture suggests a belief in the positive humanitarian consequences of the Western presence in Afghanistan, more particularly in modernity defined in economic terms as the way to material progress and prosperity.

Most of the other photo-texts in the sequence do not depict Afghans but focus entirely on the road and the machines used to make it. Processes referred to in the captions are: "feeding of the asphalt machine," "compacting of the asphalt layer," "repairing and extending culverts," and "laying an asphalt layer." These technical descriptions dehumanize the project of rebuilding the nation, stressing procedures and apparatuses, a key feature of modernity. USAID imagery not only links nation-building to modernity, but modernity is furthermore presented in its Euro-American format as a discontinuity with traditional society, and there appears to be no room for "alternative modernities" (Appadurai, 1996) with heterogeneous political, social, cultural, and economic characteristics. As in the classical social theory of Marx, Durkheim, and Weber, modernity is seen as a characteristic of Western societies, and non-Western societies are expected to become modern too. In USAID communication, the role of the West as a model of modernity is emphasized in various ways. One image in the photo-essay, for instance, portrays "U.S. Ambassador Robert Finn, behind the wheel of an asphalt laying machine," clearly emphasizing the importance of Western action and leadership. While the ambassador is placed inside modern space (the new road) in that picture, the Afghan men in Fig. 1.2 are separated from the symbols of modernity and located at its "rural" periphery.

The coupling of nation, technology, and modernity in USAID imagery, and the road as a key metaphor for nation building and progression, eclipses a great number of other elements that belong to nationhood such as its ethnic, religious, civic, and military dimensions. The USAID images omit that national homogenization has typically been a violent process of invasion, war, colonization, and discursive oppression (Conversi, 2007). The American public affairs view on nation-building in Afghanistan, of which the USAID images are a part, keeps these aspects out of sight. What the seven pictures of the USAID photo-story on the Kabul-Kandahar road do not mention, for instance, is that the road was originally built in the 1960s with money and equipment from the United States as part of their Cold War campaign to establish influence in Afghanistan. Modernization was not primarily seen in this context as a means to promote local progress and prosperity but as a weapon in the struggle for global domination. During the Soviet occupation, the road was destroyed, and later the Taliban rebuilt part of it. Since the United States entered the country and began to rebuild it once more, many have lost their lives on it. The road, then, could equally be used as a symbol of contestation, insecurity, violence, and struggle for domination in central Asia, as Kathy Gannon (2004) did in *The New Yorker*.

In USAID humanitarian imagery, the violent history of the Kabul-Kandahar road is rendered invisible. The photo-text in *Die Welt* similarly keeps violent aspects of the intervention out of sight. It omits that aid packages dropped over Afghanistan were similar in size and color to unexploded yellow cluster bombs, and a potential danger to Afghan civilians, especially children.

The examples demonstrate that photo-textual frames derive their meaning from what they include and exclude, and that photographs primarily work spatially. The meaning of "The Spirit of Berlin" is based on its upper–lower spatial division, that of the Kabul-Kandahar Road on its left–right composition. The spatial organization creates specific temporal, social, and international relations. The photograph in *Die Welt* connects Afghanistan and Germany and establishes a temporal link between the aftermath of World War II and the intervention in Afghanistan. It suggests political relationships between a liberator (the United States), oppressed populations (Germans in West Berlin, Afghans under the Taliban) and oppressors (the communists, the Nazis, and the Taliban). The USAID photograph suggests a chronological development from premodernity to modernity. Finally, the examples from *Die Welt* and the USAID belong to a singular shared transatlantic discourse. Although used in different contexts and for different purposes, both images convey a similar positive view of the intervention in Afghanistan. Both focus on the humanitarian aspect. The photograph from *Die Welt* focuses on its immaterial aspects, whereas the USAID picture emphasizes its material and technological aspects. Both phototexts frame the intervention as a constructive, humanitarian effort, eclipse its negative effects, highlight the role of the West, and relegate Afghanistan and its people to the margins. However, as I show in Chapters 3 and 4, this is only part of the story. More ambiguous visions of Afghanistan have circulated the Western world as well.

## NOTES

1. For a discussion of different perspectives on framing in communication studies, see D'Angelo (2002), Entman (1993), Reese (2001), and Scheufele (1999).

2. A useful distinction is the one between "frame advocates" who support a frame and "frame sponsors" who consciously and actively seek to promote a frame (Gamson & Modigliani, 1989). Frame sponsors are typically governments, corporations, and other organizations that may hire public relations companies, spin doctors, and other professionals to get their frame across. Governments and the military apparatus try to control what and how photographers shoot reality by offering them specific frames that easily fit into mainstream news media. Journalists may transmit, transform, or discard the sponsored frame.
3. The work of Kress and van Leeuwen (2003) suggests that visual *episodic* frames appear in two types of images. Narrative images visualize an action or event, whereas conceptual images visualize individuals in a *classificatory, analytical* or *symbolic* way. Conceptual images are classificatory when they represent depicted participants as part of the same class, analytical when they frame the depicted participants attributively, and symbolic when they suggest that the depicted participants stand for something else.
4. Although actual physical distances may vary cross-culturally, the correlation between physical and psychological distance seems to be a universal pattern grounded in practices of child nurturing and sexuality. Using Edward Hall's theory of proxemics, Kress and van Leeuwen (2003) divided psychological distance into six categories: *intimate* (face or head only), *close personal* (head and shoulders), *far personal* (from the waist up), *close social* (the whole figure), *far social* (the whole figure with space around it), and *public* (the torso of at least four or five people).
5. Several definitions of framing stress the organizational aspect. According to Reese (2001), frames are "*organizing principles* that are socially shared and persistent over time, that work symbolically to meaningfully structure the social world" (p. 11, italics added). For Goffman (1974), frames are "definitions of a situation, built up in accordance with *principals of organization* which govern events ... and our subjective involvement in them" (pp. 10-11, italics added). According to another definition, a frame is "*a central organizing idea* for news content that supplies a context and suggests what the issue is through the use of selection, emphasis, exclusion and elaboration (Tankard, Hendrickson, Silberman, Bliss, & Ghanem, 1991, cited in Tankard, 2001, pp. 100-101, italics added).
6. The blurred quality of the image brings to mind a host of other images of soldiers as tragic heroes. Focusing on this intertextual aspect, the photographers and World Press Award jury members Broomberg and Chanarin (2008) found the image evocative of the time that photojournalism was still financially rewarding and morally important.
7. All translations of newspaper captions and headlines are my own.

2

# Official U.S. Views

The conflict in Afghanistan has been a struggle with military, diplomatic, legal, financial, political, intelligence, and ideological dimensions. This chapter focuses on the ideological aspect, and the role of the United States. It considers the use of photographs in public affairs efforts by the USAID and the DoD, and in PSYOP. PSYOP messages differ in some obvious ways from public affairs messages but both share a number of discursive premises and goals too. At the start of the conflict in 2001, home and Afghan audiences alike had to be convinced that the intervention in Afghanistan served a just cause, and it was felt that national unity had to be strengthened both at home and in Afghanistan in order to overcome the threat of terrorism. PSYOP and public affairs equally served these causes.

## PHOTOGRAPHY AND PROPAGANDA

PSYOP and public affairs both fit the definition of propaganda as "a series of targeted, systematic messages disseminated through multiple channels for a prolonged period of time that offer biased opinions or perspectives through the selective use of specific, emotionally arousing, comprehensible, and aesthetically appealing techniques that circumvent scrutiny of the message to influence attitudes and beliefs" (Curnalia, 2005, p. 240). In the United States, photography has been an important instrument of propaganda since the American Civil War. The Farm Security Administration photography project (1935-1944) was a major propagandistic operation set up to promote President Franklin D. Roosevelt's New Deal, and during World War II, photography helped to create a home front. Citizens, including photographers, were expected to help in whatever way they could. Edward Steichen, who commanded the Naval Aviation Photographic Unit during World War II, used and recaptioned Farm Security Administration[1] photos in the exhibition *Road to Victory*[2] (1942) to persuade audiences to support the war effort. Later, he staged the photography exhibition *Power in the Pacific* (1945) that celebrated American might. Exhibitions remained important instruments of propaganda during the 1950s, to counter the Soviet Union's influence. Steichen, who had succeeded Beaumont Newhall as director of the Department of Photography at the Museum of Modern Art in 1947, organized *Korea* (1951) and later what remains the most popular photographic exhibition of all times, *The Family of Man* (1955). The exhibition toured worldwide and was visited by approximately 9 million people between 1955 and 1959. Many critics and photographers criticized its "universal humanism" that was believed to promote middle-class patriarchal American values.[3]

Although official support for propagandistic exhibitions gradually declined, the idea survived. In 2001, the Bureau of Educational and Cultural Affairs of the U.S. Department of State collaborated with the Museum of the City of New York and photographer Joel Meyerowitz to create an exhibition of a selection of images from the photographer's work on Ground Zero in New York.[4] *After September 11* (2002-2003) was seen at approximately 150 venues around the world, including at the Afghanistan National Gallery of Art in Kabul. Around the same time, the U.S. military began to exploit digital ways to engage audiences in its war efforts. One of its successes has been *America's Army* (2002), a computer game based on training software for soldiers. The game relies on real images to encourage people to enlist in the army and to present war as entertaining.

A comprehensive initiative to improve the effectiveness of communication in wartime was formulated by Donald Rumsfeld in the secret *Information Operations Roadmap* (Department of Defense, 2003). The document proposed the integration of public affairs, PSYOP, and public diplomacy. When the document was leaked, *The New York Times* warned that the integration would "risk blurring the traditional lines between public affairs programs in the Pentagon and military branches—whose charters call for giving truthful information to the media and the public—and the world of combat information campaigns or psychological operations" (Shanker & Schmitt, 2004, para 4). Other critics argued that the blurring of lines between public affairs, public diplomacy, and PSYOP was against the law because propagandizing the American public with PSYOP materials is prohibited under the Smith-Mundt Act, originally created to regulate propaganda in the Cold War era (The National Security Archive, 2006).[5] The *Information Operations Roadmap* was only one of many initiatives to improve information and communication strategies. In November 2001, at the time of the large-scale bombing campaign in Afghanistan, the Bush administration set up Coalition Information Centers in Washington, London, and Islamabad to improve communication about its "war on terror" and to influence the perception of the United States by foreign countries and to counter propaganda messages from the Taliban and Al-Qaeda in a rapid way.[6] Many committees, strategies, and plans followed.[7] The various initiatives, many of them short-lived, indicate that the Bush administration realized from the start that a military operation alone would not suffice.[8]

President Bush (2001) defined his concept of a "war on terror" in his famous speech before a joint session of Congress on September 20, 2001. He saw 9/11 not as a crime or a suicide bombing but as an "act of war" by "enemies of freedom" (para 12), and promised an endless global battle to fight "evil" (para 16, 22) and defend "civilization" (para 35). On the one hand, the president appealed to the entire world:

> This is not just America's fight. And what is at stake is not just America's freedom. This is the world's fight. This is civilization's fight. This is the fight of all who believe in progress and pluralism, tolerance and freedom. (para 35)

On the other hand, Bush stressed the unity of his nation. Early on in his speech, he noted that "the entire world has seen for itself the state of our Union—and it is strong" (para 5), and toward the end of it, he repeated five times: "we will come together." The president also made a distinction between the people of Afghanistan, the Taliban regime, and Al-Qaeda. The "war on terror" was defined rhetorically as a new war at three levels: global ("the world's fight"), national (the United States), and local (Afghanistan). Each level was further defined in bipolar terms. The global world was split into civilized states versus those supporting terrorism; the nation was defined as strong but threatened; and the Afghan population was split into victims and perpetrators. The plot line was that the nation was at war with the perpetrators in Afghanistan and with those who supported them, and that it had the civilized world on its side to free the victims. This official discourse of the "war on terror" represented a major shift in U.S. views on terrorism. Whereas previously attacks on Pan Am flight 103 (1988), the World Trade Center (1993), the Alfred P. Murrah building (1995), and the American embassies in Kenya and Tanzania (1998) had been framed as crimes, 9/11 was seen from the start as something that would require more than criminal investigations or retaliatory measures. The military intervention in Afghanistan was seen as the first step in this new process. In his speech, President Bush defined it as the beginning of a war that would be different from all earlier wars. Rumsfeld later confirmed that the "war on terror" would be "very, very different from World War II, Korea, Vietnam, the Gulf War, Kosovo, Bosnia—the kinds of things that people think of when they use the words 'war' or 'campaign' or 'conflict'" (Garamone, 2001, para 3).

The concept of the "war on terror" contained elements of absolute war, total war, and just war. Clausewitz (1976) used the term *absolute war* as the tendency of war "to compel our enemy to do our will" with "no logical limit to the application of that force" (pp. 75, 77). Central to *total war* is not the extent of force but the extent of the resources brought to the war. Total war involves the entire nation and its economy, each individual citizen, and not just the state and the army. It relies on workers and scientists and seeks to destroy the enemy's industry (Handel, 1986). The concept of *just war* implies, contrary to pacifism, that war is not always wrong but also, contrary to realism, that even when going to war is justified not everything in war is morally permissible. The concept has been developed into two sets of principles, one for deciding whether going to war is justified (*jus ad bellum*), the other for deciding whether conduct in war is permissible (*jus in bello*). The core *jus ad bellum* principles are legitimate authority, just cause, and right intention. According to the core *jus in bello* principles, military action must be discriminate and proportionate. The principle of discrimination, according to which a combatant is not allowed to target noncombatants, and the rule of proportionality, which states that the good consequences of a military action must outweigh the bad, is related to the doctrine of the double effect. The doctrine holds that harmful actions are acceptable when they are unintentional side effects of the promotion of a good cause. Since the Kosovo War (1999), the unintentional side effects are known as "collateral damage."[9]

The Bush administration defined the "war on terror" as an absolute war by setting neither spatial nor temporal limits to it, and Rumsfeld (2001) evoked the concept of total war when he said that the "war on terror" would involve individuals, groups, and countries, when he referred to the involvement of bankers, programmers, and customs officers to fight terrorists, and when he called on every citizen to contribute to the war effort. The civilian population was asked to accept restrictions of civil rights issued by the Department of Homeland Security and exceptional measures such as the Patriot Act of 2001 meant to enhance the surveillance and investigative capabilities of U.S. law enforcement agencies.[10] The Department of Homeland Security, which united 22 agencies on March 1, 2003, encouraged citizens to be prepared at all times for emergencies. To motivate involvement, it introduced a Homeland Security Advisory System in March 2002 to continuously remind citizens of the threat of terror.[11] The system works with several levels, from severe to low, but never reaches zero. The Homeland Security Advisory System thus institutionalized fear even though President Bush described the "war on terror" as a fight against fear.[12]

The state sought to motivate the public through a representation of the "war on terror" as a liberation struggle too, conducted "to free the victims." The United States went to war not simply because it was attacked, but because the country needed to liberate an oppressed people. The National Strategy for Combating Terrorism (2003) called the long-term approach of "the war on terror" the advancement of freedom and human dignity through democratic governance, the rule of law, and the facilitation of economic development. Freedom became central in the naming of the intervention in Afghanistan when it was turned from Operation Infinite Justice into Operation Enduring Freedom in 2001. The freedom and justice concepts defined the intervention in Afghanistan in accordance with the *jus ad bellum* rules of just war.

Although many qualified the initial intervention in Afghanistan in 2001 as justified, others found that the intervention did not meet all of the criteria of just war. Shalom (2009) argued that the intervention failed to meet the criteria of last resort, legitimate authority, and proportionality. For Shaw (2002), the intervention in Afghanistan was not a just war because the disparity between the Afghan civilian death toll and the American military death toll was enormous. Even though the number of civilian casualties in Afghanistan remained small by comparison with Vietnam and Korea, it appeared large when compared with the number of Western military casualties. Others argued that national interest theory applied, not just-war theory. Beck (2005) called the combination of humanitarianism and national interests the "idealism-realism synthesis" (p. 17). From a realist perspective, states operate on the basis of power and self-interest and humanitarian preoccupations cannot be a reason for intervention, but the double standard was welcomed. Robert Cooper (2002), a senior British diplomat and advisor to British Prime Minister Tony Blair, called it "a new kind of imperialism" that would bring human rights, cosmopolitan values, order, and organization (para 14), and the National Security Strategy of the United States (2002) regarded this "new imperialism" as a guarantee for sustainable peace on a global scale.

The confluence of human rights, democracy, and military action was institutionalized in Afghanistan in Provincial Reconstruction Teams made up of military and civilian personnel. First called Joint Regional Teams, these units were proposed by the United States in

November 2002 to facilitate the transition from security operations to reconstruction. Many nongovernmental humanitarian agencies in Afghanistan received this blurring of civilian and military identities as a threat to their independence and security especially after the murder of five Médicins Sans Frontières workers in June 2004 (Save the Children, 2004).

Although public affairs imagery of the DoD generally stresses military aspects, it has been marked by the double standard of humanitarianism and militarism. To some extent, DoD public affairs pictures are a visual translation of the new imperialist discourse. Many images stress the technological might of the military but other pictures focus on humanitarian assistance. In both cases, the role of the soldier is highlighted. The same applies to PSYOP imagery and to some degree to representations in the press.

## A SOLDIER'S STORY

According to its information policy, the DoD seeks to provide the public with accurate information (Department of Defense, 2008). Restrictions and omissions are explained as necessary for legal reasons or for the protection of ongoing operations and security. But the DoD images omit and emphasize particular information for public affairs purposes. DoD images are intended to legitimate U.S. military activities, not to provide accurate information on Afghanistan and its activities. Afghanistan is peripheral to the DoD story, and the focus in DoD imagery falls entirely on the United States and its soldiers. DoD Web sites and their images offer a very limited perspective on the conflict in Afghanistan. They neither show evidence of the damage caused by military operations nor do they attempt to shed light on the structural and cultural roots of the conflict. As is to be expected in a public affairs context, the pictures present a sanitized view of war, displaying no direct violence, virtually no damage, and almost no wounded or dead. In my sample, I found four images of wounded Pashtun combatants, and one series of images of a ceremony for a fallen soldier. Apart from some photographs of detainees at Guantanomo Bay, depicted from the back, the enemy remains mostly invisible and is implied only.

The first image that was published in the photo gallery of Operation Enduring Freedom on the DoD's DefenseLink Web site depicted the launch of a cruise missile (Cosgrove, 2001). DoD images frequently show missiles and bombing technologies, but the actual explosions on the ground and their consequences are rarely revealed. The bombings are referred to in some captions, where they are described in just and limited war[13] terminology as, for example, "carefully targeted actions," "designed to disrupt the use of Afghanistan as a base for terrorist operations and to attack the military capability of the Taliban regime" (see Fig. 2.1). The wording avoids direct association of war with the killing and maiming of individual human beings. Combat-related action shots often show preparations for war such as cleaning weaponry, loading ordnance, and preparing aircraft, but not the actual fighting. Many images focus on other "logistic" activities such as the movement of troops and vehicles in the context of combat, or search and surveillance activities. Some images focus on preparing and maintaining the body through physical exercise and training, and other pictures depict logistic processes such as setting up camp, disposing of ordnance, destroying mines, constructing buildings, and preparing landing strips. This emphasis on the preparatory stages of combat, the movement of troops and logistic activities can be read as an attempt to satisfy the public's desire for war imagery without revealing the horror of battle and the consequences of war. Alternative visual records of the war, such as those collected by the Revolutionary Association of the Women of Afghanistan (n.d.), often draw attention to the consequences. They display damaged buildings and mutilated civilian bodies, especially those of women and children.

**Fig. 2.1. "The night launch of a Navy F-14 Tomcat". David A. Laviolette (2001)**

"The night launch of a Navy F-14 Tomcat becomes a trail of steam and light silhouetting the flight deck crewmen on the USS Enterprise (CVN 65) on Oct. 7, 2001. The Enterprise is launching aircraft for strike missions against al Qaeda terrorist training camps and military installations of the Taliban regime in Afghanistan as part of Operation Enduring Freedom. The carefully targeted actions are designed to disrupt the use of Afghanistan as a base for terrorist operations and to attack the military capability of the Taliban regime."

The framing of action by DoD military photographers partly resembles that by early war photographers, when for technical reasons the actual combat could not be captured easily. American Civil War photographers focused on the preparations for war and daily camp life. But these early photographers captured the devastations of war as well, whereas DoD photographs hardly show any ravage or suffering. Instead, military photographers frequently use color, blurred motion, nightlight, and *contre-jour* to render war and military scenes abstract, aesthetically pleasing, and spectacular. Although military photographers often employ the codes of the photojournalistic genre to simulate a morally neutral frame, they regularly opt for a more aestheticizing approach too that celebrates warfare as beautiful. Figure 2.1 is an example. The blurring in the photograph in Fig. 2.1 has the additional effect that it focuses attention on the motion or speed of the action, and thus accentuates the power of technology.

Military photographs have represented the war in Afghanistan as a reassuringly ordered and bloodless conflict, with U.S. troops in full control. Photographs of search and surveillance activities present the viewer a panoptical range of surveillance activities that extends from the close examination of mosquitoes, Afghan bodies, and livestock to the inspection of houses and the Afghan territory. Images of security activities include those of troops inspecting the land, the villages, the buildings and construction sites, the vehicles, and weapon cache sites without any apparent calamity. The moments selected to represent the action rarely involve the local population, the Taliban, or members of Al-Qaeda. The photographs feature U.S. soldiers running through a village or entering the door of a village home. Action scenes either tell the story of the soldiers *in medias res* or *ab initio*, as in the pictures of the preparatory or initial stages of combat. They work ideologically because they keep the other moments of the sequential action out of sight, in particular the consequences of the action. Filmmakers have represented surveillance activities in Afghanistan more fully, and in different ways. A Norwegian documentary made by John Sullivan (2003), for example, shows how during raids, private rooms are entered, children start crying, and men are hooded, imprisoned, and threatened. The documentary explains that Afghan civilians are twice victimized: They may be forced to hide illegal weapons, and they become victims again when U.S. soldiers find the weaponry in their houses. In "Scenes of Afghanistan after the U.S. bombing," an online collection of eight files with a vast number of photographs, Marc Herold (2004-2007), a development economist and activist at the University of New Hampshire, sought to counter-frame the searching of private homes, as an ongoing disturbance for Afghan people. One strategy employed by Herold is the juxtaposition of images "intended to stimulate thought about the connections between war as a contest and spectacle, the professional sports metaphor in war, and the female imagery often harnessed to support the war effort" (Herold, 2004-2007, para 3, file 1). He also reframed original DoD images by recaptioning them to stress the killing in the bombing campaign and the harassment of civilians during search operations.

When military photographers rely on the aesthetic familiarity of photojournalism to create a sense of objectivity and neutrality, this goal often is pursued with seemingly descriptive captions. DoD captions provide facts and focus on the precise naming of devices, as in "Aviation ordnancemen of the VFA-131 'Wildcats' on board the USS John F. Kennedy load an AIM-9M Sidewinder air-to-air missile on an F-18C Hornet in preparation for its next combat mission" (Hampshire, 2002) or "The 'Shooter' gives the signal to launch as an F-14D Tomcat from Fighter Squadron 213 is pulled down the flight deck of the aircraft carrier USS Carl Vinson by a steam-powered catapult" (Messier, 2001). The consistent use of technical jargon and abbreviations and the frequent framing of vehicles and other technologies as actors[14] convey fetishism.[15] Techno-fetishism also is apparent in photo-sequence titles such as "The B-52s: Big and tough" and "Fast Jets and Sharp-Dressed Men" (Department of Defense, 2001).

Concrete captions sometimes contrast with more abstract visual representations, as in Fig. 2.1. The image creates a view of the war as an abstract visual spectacle, and the second half of the caption frames the war as a precise technical operation.

DoD texts generally describe warfare in dehumanized, technical terms, or occasionally draw attention away from the military action to the act of photography and the role of the photographer. Soldier-photographer Johnny Bivera comments on an image reproduced in the next chapter (Fig. 3.1) as follows:

> Combat Camera is all about action. When you train and document this sort of thing as much as we do, knowing when or where the action is going to occur can only come from experience. I like to explain to people, and definitely our young photographers, that once you hone your skill as a shooter, the actual picture-taking part of the job is what's easy. Getting from point A to B with all your gear, shoot the job and return unharmed, that's what's hard. (Bivera, 2002)

The text evokes the mythical representation of the photographer as a brave and adventurous hero who partakes in the action, with the actual picture-taking presented as almost incidental to the danger and action shared by photographer and soldier. The use of an all-inclusive "we" and the reference to "our young photographers" suggest an atmosphere of comradeship and shared physical ordeals. In other photo-texts, comradeship is combined with a focus on the physical capabilities of soldiers. Strong diagonal lines and low angles are frequently used to enhance the view of soldiers as dynamic, virile, masculine, and powerful comrades. Many of the DoD photographs also depict soldiers as technological operators through an association with technological apparatuses. They are shown cleaning their weapons, carrying a bomb, preparing an aircraft, moving around in a vehicle, or testing equipment. The symbiosis of man and machine is enhanced in several photographs through stripping away the wider environment, by reducing the color range so that man and machine look more similar, through silhouetting both, or through a combination of focal length, tight composition, and camera angle. In Haraway (1991), the cyborg challenges modernism and established power relations by questioning the radical separation between humans and technology. In DoD images, the man–machine symbiosis merely symbolizes the modern myth of control through technology, and encourages viewers to identify with the technological apparatus of the war and to feel as powerful.

A belief in the power of technology seems to encourage many military photographers to include and focus on the technological means of action. Apart from binoculars and other surveillance equipment, these include weaponry and means of military transport. Soldiers are represented as techno-warriors, equipped to fight a clean and rational battle. But technological devices regularly become actors too. Pictures focus on helicopters landing, Humvees on the road, and aircraft in midair. These images do not depict any human being, or reduce the presence of human beings to tiny spots.

These photographs, that "objectify" the war, have been published frequently in the press as well. They were especially popular in the period of the build-up for war, prior to the start of the actual attack in 2001, when action shots could not yet be made. Instead of glorifying soldiers at the front line—an older theme in war photography—images of techno-actors stress technological supremacy and glorify the technology of war, a theme that has gained prominence since the introduction of military aircraft during World War I.

In DoD images of humanitarian warfare and development, there is a strong element of control: Soldiers control the bodies of Afghans through the provision of medication and medical examinations; they control the schooling of Afghans through the provision of textbooks and the construction of schools; they control their nutrition and regulate the provision of goods. The emphasis on the physical body in several pictures and captions about the medical treatment of the Afghans, their livestock, and their hygiene reveals a neocolonial desire to discipline the indigenous body. The old European colonial powers were also obsessed with taking possession of the body of the colonized subject: with dressing it, washing it, examining it, healing it. In the colonial psyche, it is not just the land and the mind but the body of the colonized subject that has to be conquered (McClintock, 1995). This colonial belief continues to speak through a number of DoD photographs, and, as I discuss later, through press images as well. Consistent with the colonial perspective, DoD imagery frequently represents the Afghan landscape as an empty territory without an indigenous population and with only foreign troops moving through. Photographs of soldiers looking through binoculars depict the first step of taking possession of the land. Settings are often literally obscured by placing U.S. soldiers or technological devices in the foreground, by applying shallow depth-of-field to blur the background setting, or by shooting *contre-jour*. Monochromatic colors, haze, night photography, and dust from helicopters landing or arriving frequently give settings an abstract quality. Similar abstractions from the human geography are realized by photographs of maps and by remote aerial photographs that reveal little or no detail. The hiding or abstraction of landscape features may serve a military-tactical purpose when information on concrete locations has to be kept secret, but they also bear ideological meaning. Photographs of empty landscapes often suggest that the displayed territory is readily available and can be freely entered, a perspective reversed in Paul Seawright's *Hidden* (2003), which focuses on the contamination of the territory (see Chapter 4).

The military landscape generally remains empty, but textbooks, shovels, stethoscopes, paintbrushes, and construction tools fill up the space of humanitarian assistance. In humanitarian scenes, U.S. soldiers appear as negotiators partaking in meetings, as recipients who receive gifts of gratitude from the Afghan population, and as common people, "like us," engaged in everyday activities. They are shown handing out clothing, toys, school supplies, hygiene or other items to local civilians. The soldiers are represented as actor-caretakers, and Afghans as their patients and beneficiaries.[16] Afghans are depicted as grateful, in blue-collar jobs contributing to the rebuilding project, or in a servant role.

As is to be expected in a public affairs context, DoD images do not focus on the disruptive consequences of war for the civilian population. Images instead frequently draw attention to the positive relations between soldiers and Afghan people, mostly children. DoD images depict soldiers as surrogate fathers, shaking hands with children, playing with them, and providing them with goods and medical care. These pictures suggest that international relations are analogous to family relations: The United States is the father, Afghanistan the child.

In accordance with 19th-century colonial ideas, the colony is not just represented as infantile but also as feminine. Afghan women typically are depicted in a context of learning and medical care, whereas female U.S. soldiers are often portrayed in the stereotypical nursing role. The frequent association of Afghan women with formal education seems to suggest that they are ignorant and in need of education. Afghan women are hardly ever presented as agents of their own destiny, and DoD photographs also reveal patriarchy at home. In DoD homecoming pictures, American women are awaiting their men, who have gone out "to protect them." The women often are accompanied by children, and thus portrayed as "mothers," whereas the returning soldiers are represented as heroes who look after the security of the nation both at home and abroad. Their unification is a ritual typical of a "participatory culture" with "a deep passion for unity" (Hallin, 1997, p. 208), one that fuses the military, the patriotic, and the personal. At homecoming ceremonies, the military photographer acts as a family photographer in the Bourdieuzian sense of the word (1990), creating images that reaffirm the unity of the nuclear family and the *nation-as-family*.

In Fig. 2.2 the homecoming ritual is presented in a digital version of the traditional family photo album, complete with photo corners and handwritten texts. The photograph of reunion is located underneath a stars and stripes banner. Symbols of the private (the photo album, the family portrait) are mixed with the symbol of the nation (the flag). The presence of the American flag reinforces the idea that the represented participants are not mere private individuals but part of the larger national collectivity. As in situations of total war, the family no longer resides in its private home but has entered the public and military space. In accordance with the ideal–real logic of vertical composition in Western culture (Kress & van Leeuwen, 2003), the real family below is made subservient to the ideal nation above, which is represented by the Defend America masthead. The self is not seen in these pictures as an autonomous and independent being but represented intersubjectively as part of the larger body politic.

Soldiers in individual portraits also are represented as part of a larger social group. They are defined as members of the military community through their attributes: their weapons and uniform. Captions further frame them as part of the national and local collectivity. For example, a caption accompanying a portrait of "Army Spc. Jeffrey E. Reen" (Department of Defense, 2002) defines the individual by focusing on "his participation in Operation Anaconda and Operation Harpoon in Afghanistan," his membership in "Alpha Company, 4th Battalion, 31st Infantry Regiment, 10th Mountain Division, Fort Drum, N.Y.," and his place of civilian residence in the United States, "Easton, Mass." A dominant frame in DoD imagery, then, is *collectivism* or *unity*. In Hofstede's (2003) study of cultural dimensions, the United States has individualism as its highest dimension, along with only a handful of other countries. The DoD images do not necessarily contradict this observation. They reflect the military culture, function as an affirmation of the military community, and through this also reproduce the national collectivity. DoD images of cooperation, comradeship, and reunion express a

**Fig. 2.2. "Operation Tribute to Freedom: Homecoming Photographs." Photo: Susan C. Milton (2003)**

"Family and friends watch from the pier and wait to welcome home the crew of USS Donald Cook as it returns to its homeport of Naval Station Norfolk, Va. Donald Cook returned to Norfolk May 23 after a 6-month deployment with USS Harry S. Truman Carrier Strike Force in support of Operations Enduring Freedom and Iraqi Freedom."

belief in strength through unity. The pictures convey the motto of the United States, *E pluribus unum* (*Out of Many, One*), the idea of social cohesion and unitary, collective experience.

In some photo-texts, the focus is more on individual leadership. Apart from pictures of elite political and military leaders at the beginning of the war, several photographs portray individual soldiers in their role of leader, particularly in scenes depicting negotiations with Afghans. These images of individual leadership constitute a second *master* frame in DoD public affairs. Most of the DoD public affairs photographs, however, stress leadership through technological supremacy or focus on surveillance activities, presenting the viewer with an army that is ready and in control. Technology is central to a *managerial* framing of war. It is further suggested that the intervention can be successfully managed provided the nation is united and the leadership appropriate. The *unity* and *leadership* frames are thus closely related to the *managerial* frame. Unity and leadership are presented as important preconditions for bringing the military and humanitarian mission in Afghanistan to a successful end.

## PAPER BULLETS

The themes of technological control and military supremacy reappear in PSYOP. PSYOP are "the planned use of communications to influence human attitudes and behaviour, to create in target groups, behaviour, emotions, and attitudes that support the attainment of national objectives" (Narula, 2004, p. 184). The term was first used in the context of the surrender of Japan in 1945 but the practice itself is probably as old as war itself. Today, PSYOP are one of the core activities of U.S. Information Operations (Lamb, 2004) along with computer network operations, electronic warfare, operations security, and military deception.[17] PSYOP are seen as "a vital part of the broad range of U.S. diplomatic, informational, military, and economic activities" (Joint Chiefs of Staff, 2003, p. ix). Since 2001, their role seems to be on the increase. Reported plans to expand the 4th Psychological Operations Group at Fort Bragg, from 1,300 troops to about 2,300 by the year 2011, are one indication of their importance (GlobalSecurity.org, n.d., para 1). Another is the number of PSYOP leaflets that were dropped in recent conflicts. During the first Gulf War, the United States dropped around 29 million leaflets (Pfeffer, 2003); in Afghanistan the number rose to 84 million flyers in the first year of the conflict alone (Friedman, 2002).

PSYOP messages can be targeted at various audiences and can serve multiple and often very concrete purposes. They are set up to warn civilians of mines and food droppings, to provide them with specific instructions, or to promise the population something; they also are used to legitimate military actions, to counter enemy propaganda, and to invite cooperation. Messages targeted at enemy forces usually seek to demoralize and divide. They also may strive to create allegiances. At the start of the intervention in Afghanistan, general PSYOP objectives were to isolate the Taliban and Al-Qaeda from support, to increase the legitimacy of the operations, and to undermine enemy morale (Lamb, 2005). But PSYOP messages were also used to garner support for elections and the government.

The PSYOP strategy in Afghanistan has encompassed military, intelligence, political, and ideological goals, and these have been pursued through different channels. Radio, the most important form of mass media in Afghanistan, has been used widely. During the initial intervention in 2001, military aircraft propaganda broadcasted spoken messages and music, which the Taliban had forbidden. Small battery-powered radios were distributed on the ground to receive the U.S. programs, and the Voice of the Sharia radio station of the Taliban was bombed to avoid counter-propaganda. PSYOP efforts were later extended with a ground-based radio station, Radio Malumat, and a newspaper (Sammons, 2004). In addition to these more traditional activities, PSYOP specialists engaged in "propaganda of the deed" when they solicited support from local leaders and supplied schools and hospitals.

PSYOP preparations started on September 12, 2001, one day after the attacks on the World Trade Center and the Pentagon, with a target audience analysis of the Afghan populace, the Taliban, and Al-Qaeda (Lamb, 2005). A Joint Psychological Operations Task Force was set up by the end of September to support the military campaign that started on October 7, 2001. This task force produced the PSYOP leaflets and radio scripts for use in Afghanistan at Fort Bragg in the United States.

The first paper bullets, as PSYOP leaflets are sometimes called, were dropped a week after combat operations began, in the eastern, western, and northwestern parts of the country. They were the first in a series of approximately 100 different types directed to Afghan civilians, the Taliban, or Al-Qaeda members. The leaflets neither concealed their origins nor their objectives. They were classified by the military as very low secret and quickly became public after being dropped (Pfeffer, 2003). U.S. Central Command made some of the leaflets available itself, and the press published several.

The leaflets were usually about 3 x 6 inches. The design and illustration work had an amateurish look. Most of the flyers contained two or more images, with short texts in Pashto and Dari, to be read from left to right. Temporality and causality were expressed by the juxtaposition of images, and sometimes by an additional arrow. Many leaflets carried symbols of peace and unification (dove, country map, flag), prosperity and profit (money, thread, and rug), or evil (spider, dog). The color green, symbolic of Islam and associated with paradise, reappears frequently.

At first, PSYOP tried to divide the Taliban, and to break allegiance of Taliban members to Al-Qaeda with leaflets that portrayed Al-Qaeda members as foreigners who manipulated the Taliban. Symbols and cartoons appear frequently on these leaflets. In one cartoon image, bin Laden is portrayed as a chess player moving Taliban members on a chessboard; in another bin Laden is a spider holding members of the Taliban in his web. Later, PSYOP messages more explicitly incited defection and tried to demoralize the enemy by stressing American strength through surveillance and combat technology. These leaflets illustrated the potential effects of U.S. bombing on Al-Qaeda members, or depicted crosshairs to suggest panoptical control.

PSYOP messages targeted at civilians were not limited to warnings against mines, grenades, and bombs, or to instructions for receiving propaganda radio or using food packages. Several messages targeted at civilians represented the Taliban and Al-Qaeda as an evil collective. Mullah Muhammad Omar and bin Laden were featured in these messages as murderers, cowards, and traitors. Both leaders were associated with the attributes of abjection: corpses, drugs, and impure bodies. Building on Douglas' *Purity and Danger* (2002), originally published in 1966, Kristeva (1982) defined the abject as a threat to normal distinctions between pure and impure bodies, and to established identities and communities more generally. To uphold traditional community, is to expel the abject that confuses conventional distinctions, according to Kristeva. The corpse is the exemplary abject for her. It is neither subject nor object, and marks the end of identity and community. The text of a leaflet with a cartoon of bin Laden on a pile of dead bodies associated the Al-Qaeda leader with this extinction of community: "Osama bin Laden sacrifices the Afghan people for his own pride. He used the Taliban to exterminate whole communities opposed to his fanaticism." Associations between harm done to the physical body ("drugs") and the end of community ("murder") were used to portray Mullah Muhammad Omar as well. The text of one leaflet explained: "He made millions of dollars selling evil drugs to Muslims. He did not use his fortune to help the Pashtun people. He used his fortune to help Osama bin Laden murder innocent civilians. Now Omar the coward and traitor hides in safety and comfort while Pashtuns suffer."

One photomontage (Fig. 2.3) featured bin Laden as a criminal colonial *évolué*, without a beard and in Western clothes. The image clearly was meant to evoke repulsion and disgust in a way that abjects do, even though the traditional attributes of abjection—dirt, decay, and other impurities—are literally and totally absent in the image. Figure 2.3 illustrates that impurity is not an inherent characteristic but determined by context. Bin Laden is cleanly shaven and well dressed, but this precisely signals his criminal abject status. The shaven skin and Western dress suggest a transgression of boundaries, and exclude him from his own community. They are signs of bin Laden's status as a traitor who threatens the normal order. Implicit is the moral warning that association with bin Laden makes one a traitor and criminal too, and will lead to expulsion from the community.

In some PSYOP messages, the texts more directly incited the population to act against "the foreign terrorists," and reward money was depicted to motivate participation. For example, during Operation Anaconda in March 2002, the United States used a leaflet in the form of a banknote in the Shah-e Kot Valley and Gardez in Paktia Province. The text on the back addressed its audience with an all-inclusive "us": "Dear countrymen: The Al-Qaeda terrorists are our enemy. They are

**Fig. 2.3. "Osama bin Laden, the murderer and coward, has abandoned you."** (PSYOP leaflet TF11RP03)

the enemy of your independence and freedom. Come on. Let us find their most secret hiding places. Search them out and inform the intelligence service of the province and get the big prize." Some of these leaflets were reminiscent of the Wild West reward notices, and President Bush's words that he wanted Osama bin Laden "dead or alive." In other PSYOP messages, the reward theme was visualized by means of banknotes too, and with images of bin Laden behind prison bars. The text of one reward leaflet explained: "Get wealth and power beyond your dreams—help the anti-Taliban force to rid Afghanistan of murderers and terrorists." Several images drew attention to the violence done by the Taliban or Al-Qaeda, apparently in order to break allegiance and support. One caption of a photo of a man whipping a woman addressed a male audience: "Is this the future you want for your women and children?"[18]

One striking feature of PSYOP in Afghanistan is that many messages were ideological and soft, with relatively few leaflets carrying pictures of dead and mutilated enemies. In the first years, PSYOP used ideological messages to convince the population of the need for stability and unity, and to garner support for Hamid Karzai. PSYOP messages needed to persuade the inhabitants of Afghanistan that the coalition forces were liberators, and that the intervention should not be seen as an attack on them or their country. Messages distributed during the initial intervention represented Afghan civilians as beneficiaries of the intervention in an attempt to create goodwill and in order to motivate the population to support the nation-building process and the government. Leaflets tried to win the hearts and minds of the Afghan population through displaying empathy with Muslims. These messages emphasized the prosperity and freedom of Muslims in the United

States, stressed that food aid complied with Islamic food prescriptions, and included congratulations on Islamic holidays.

The humanitarianism of the United States was opposed to the violence of the Taliban. One side of a leaflet, with the text "America has provided over $170 million in aid to Afghanistan," depicted Afghans unloading sacks of food with U.S. labels, and smiling children; the reverse side, with the text "This is what the Taliban had done," had images of destruction, wounded Afghans, and a man training a rifle on an Afghan woman. Around the same time, this double message of oppression and humanitarianism was targeted by public affairs at domestic audiences, and it entered the press as well. The humanitarian message promised a better material ("money," "wealth") and immaterial ("freedom from oppression") future. National unity and international cooperation were promoted as essential preconditions for material and immaterial gains.

The first set of approximately 350,000 leaflets distributed in Afghanistan depicted a U.S. soldier shaking hands with an Afghan, accompanied by the words "The partnership of nations is here to help." The handshake was one of several Western symbols used in PSYOP messaging to express national unity, international agreement, or ethnic solidarity. One leaflet with a globe as its major symbol had a similar text as the handshake flyer, framing Afghanistan similarly in an international context: "Afghanistan. The partnership of nations is here to assist the people of Afghanistan." Another flyer with a flag, a map, a handshake, and an American and an Afghan family stressed the need for unity too. The text of the leaflet consisted of three words: "Friendship, America, Afghanistan."

Two flyers with a white dove of peace on a map of Afghanistan and with images of reconstruction (a building site) and prosperity (spools of thread and a rug) presented national unity as a precondition for peace defined as material prosperity. One leaflet with a map of Afghanistan and four portraits apparently meant to represent the nation used prosperity and freedom as a *motivational* frame: "A united Afghanistan = peace prosperity. The future of Afghanistan depends on your support of the new government. A new government offers new freedoms." The portraits suggested opportunities for women, respect for the elderly, the values of the nuclear family, and freedom. Another leaflet with images of a girl, smiling children, and two boys called on the reader to support the government: "Help bring back happiness to Afghanistan. Supporting your new government offers a brighter future for you and your children." Once again, the text introduced immaterial benefits ("happiness") and material prosperity (one of the boys in the picture has cash protruding from his pocket) as *motivational* frames.

A few leaflets, used with the intention to build international rapport and understanding, focused on the 9/11 attacks and on members of the

Muslim community in the United States. According to Friedman (2002), U.S. Central Command worried about the credibility of flyers depicting the burning World Trade Center towers because Afghans might not have seen them. Flyers with the burning towers were therefore only distributed near the end of the invasion of 2001. On one leaflet, the United States was presented as the country of the free. Focusing on religious freedom and counter-framing Al-Qaeda propaganda about the United States, one side of the leaflet had an image of the Islamic Center of Long Island mosque. Two Muslim women were depicted, one of them holding a candle and the U.S. flag. The text stated affirmatively: "Muslims in the United States worship freely."[19] The reverse side with the mosque's interior introduced the evidence: "There are more than 7 million Muslims and 1,200 mosques in the U.S."

The PSYOP visualizations of peace as freedom and prosperity are consistent with popular Western peace concepts. According to Rinehart (1989), these peace concepts are materialistic (related to prosperity), international (related to a relation between nations), and external (socially structured rather than interactional or a state of mind). For Rinehart, they are inseparable from worldviews with a basic orientation of fear: Humans are seen as fundamentally conflictual in a Hobbesian way, although they can organize themselves to behave nonconflictually.

## THE WHITE MAN'S BURDEN

Several PSYOP ideas, such as *freedom*, *prosperity*, and *cooperation*, return in USAID imagery. The USAID is the principal state agency for foreign assistance in the United States. It was established by the Kennedy administration in 1961 to administer economic assistance programs, but its origins date back to the Marshall Plan. In January 2002, the USAID reopened its mission in Afghanistan after a period of downsizing as a result of considerable reduction in foreign aid spending by the U.S. Congress. In accordance with its public affairs mission, the idea of the intervention as a constructive humanitarian effort has dominated the visualization of the USAID in Afghanistan. USAID photographs focus on restoration work, formal schooling, health care, economic development, and democratization. Unlike the public affairs photographs of the DoD, which depict actions *in medias res* or *ab initio*, the USAID images tend to draw attention to the *consequences and results* of USAID interventions but only to those that the USAID values positively. Between 2002 and early 2006, 114 USAID-funded workers were killed, 130 were seriously injured or disabled, and another 33 staff were killed in accidents (Kunder, 2006), but such aspects are kept out of sight.

To promote its new mission, the USAID cosponsored an exhibition called *Out of the Shadows* (2004). The exhibition title suggests an end to oppression and obscurity and the beginning of a new era for Afghanistan, and a USAID press release described it as "saluting the strength and spirit of the women and children of Afghanistan. At home, in school, at work, and in leadership, *Out of the Shadows* tells the story of their transforming lives today" (United States Agency for International Development, 2004, para 1).

Just as the *After September 11* exhibition mentioned earlier was meant to support the official 9/11 version, *Out of the Shadows* was constructed to spread USAID's official perspective on development and nation-building in Afghanistan. The text of the online version of the exhibit explains USAID's mission in Afghanistan as: "to rebuild what had been lost, restore what had been damaged, and put the country on the path to a peaceful and prosperous future" (United States Agency for International Development, n.d., para 4).[20]

Broadly speaking, nation-building encompasses the provision of security, state-building, and peace-building. *Peace-building* has been defined as a comprehensive approach that integrates development (e.g., Duffield, 2001; Junne & Verkoren, 2004); *security operations* in Afghanistan have involved traditional peacekeeping as well as combat (Rupp, 2006); and *state-building* includes rebuilding the structures of government and institutions, and imposing the rule of law.

The USAID photographs that I discuss here relate to what the organization considered the first stage of nation-building in Afghanistan: emergency relief and basic services. The organization legitimates its Web site as a historical record of this work while it actually serves a public affairs function and is meant to provide the general public with evidence of the USAID's efforts in Afghanistan. It obviously strives to represent these as successful. Illustrative of the gap between military and humanitarian spending, USAID public affairs are a limited enterprise when compared with DoD public affairs. Whereas the DoD relies on professionally trained photographers to create its public image, the USAID encourages its employees to submit success stories and related photographs for publication through an initiative called Telling Our Story. USAID staff or staff members of organizations contracted by it deliver most of the images. As its title suggests, Telling Our Story is not about the local population, but about the USAID. USAID guidelines for the Telling Our Story initiative mention that photographs make stories more powerful. They are valued as important means to demonstrate the impact of the USAID mission. Process- and participant-oriented framing is encouraged, apparently in order to produce the aesthetic familiarity of popular photojournalism, but many stories do not meet the requirements.

Although Telling Our Story wants to engage all USAID staff in its public communication, no efforts seem to have been made to implicate Afghan participants. All the USAID stories that I looked at have in common that they speak *for* the targeted beneficiaries implicated in the organization's aid and development processes. All are third-person narratives. The USAID does not suggest any interview model for Telling Our Story, and very few articles use quotes even though guidelines encourage their use. As a result, communication by the USAID is basically a monologue about aid and development. The origins of development, as is to be expected in this context, are mostly seen as foreign. The emphasis is on development *for* Afghans, a neocolonial vision that cannot be separated from notions of power and domination. The theme of development *by* Afghans is hardly explored. USAID discourse reveals a needs-based approach to development and a lack of consideration of existing traditions and assets. Captions typically refer to the USAID, and the organization portrays itself frequently in terms of how much money it has spent on a program. Pictures are included for their evidentiary status, as proof that the money has been spent well.

Developmental circumstances in USAID photo-texts are represented in four major ways. First, in a number of photographs, the developmental process is largely *dehumanized* through an emphasis on the means for development. Examples are photographs of machines used for the reconstruction of Highway 1 (see Fig. 1.1), of radios for receiving information on humanitarian initiatives, and of construction sites and communication infrastructure. These images are analogous to those with a focus on technology and weaponry found in DoD public affairs. In both cases, Afghans are excluded and the intervention in Afghanistan is defined as a predominantly technical matter. Second, development is *institutionalized* through a focus on schools, offices, government buildings, and refugee camps. These images suggest control through the state apparatuses and are analogous to DoD pictures depicting humanitarian operations. Third, development is seen as a restoration process: A handful of photographs of Afghanistan focus on the *consequences* of violence, what needs to be repaired, and the repair itself. Using juxtaposition, images depict the ruined landscape, ransacked buildings and damaged equipment on the left, and their repair on the right. Finally, the developmental circumstances are represented as *exotic*. Some photographs are meant to evoke the beauty of the country: a mosque against a colorful sky, women in their colorful traditional dresses, the smile of an Afghan girl. These images suggest a neocolonial and patriarchal frame in which Afghanistan appears as beautiful and feminine.

In narrative images of developmental processes, Afghans are enjoying the benefits of a USAID program. These photographs depict children going to school, women receiving medical treatment, children collecting water from a newly installed pump, Afghans harvesting a crop, men driving their tractor on a newly laid road, Afghans working on a computer, teachers working in the classroom, and so on. Other narrative images depict Afghans as auxiliary agents who carry out supportive tasks in a participatory development context, sometimes supervised by a foreigner. The USAID also regularly depicts Afghans in situations of nonparticipation, as bystanders or beneficiaries in a "development for them" situation (see Fig. 1.1).

In analytical images with a focus on participants, Afghans often are depicted as recipients of goods. Children and teachers, for example, are represented with their textbooks. Other analytical photo-texts create an exotic frame by focusing on the traditional dress of the depicted participants. Still other images focus on the facial expression and display a range of emotions, regularly those that Westerners associate with children, such as joy. Classificatory images of Afghans often are symmetrical, representing the population as a single collective entity. One photograph depicting children carrying Afghan and U.S. flags suggests a unity of nations, and a subordination of individuality to political belonging. It is a theme found in DoD public affairs and PSYOP too. Hierarchical classificatory images of Afghans mostly are taken in schools, with teacher and pupils, or in the context of an economic activity.

Of the USAID images that I examined, around 10% featured children, either as individuals or in groups. Children in the USAID images are defined circumstantially, often by portraying them in school, or attributively through a focus on their smile or dress. The first suggests ignorance (they need to learn), the second innocence and exoticism (they look harmless and beautiful). Photographs of children often have been used by aid organizations to create empathy among the public but they may also contribute to what Moeller (1999) called "compassion fatigue." They also tend to evoke the neocolonial image of a "third world" that is not yet grown up. This nation-as-child symbolism can be traced back to the old cultural colonial analogy, which holds that "primitive cultures" correspond to the infantile stages in human development and "developed cultures" to its adult stages. This analogy denies the represented nations a historical and cultural tradition of their own, and this is further stressed when children as symbols of the nation are portrayed in a learning context, as many of the USAID images do. The representations tend to suggest that indigenous people are unable to understand their own situation and to cope with it by themselves. Modern industrial states are presented within this context as grown-up adults who can tell the child-nations what to do. The grown-up states do not only possess the ability to decide what is good and what is bad, they also have the moral duty to take care of the children and to elevate them to a higher level of civilization and progress.

The *nation-as-child* frame allows for a reasoning that ultimately legitimates military intervention because the grown-up parent has the moral duty to put the child on the proper path and holds the right to punish it when it misbehaves.

Photographs of the Afghan nation as a child have appeared in prominent positions, for example, on the cover of *Afghanistan Reborn* (United States Agency for International Development, 2004), a magazine that refers to infancy in its title. The cover image depicts a child on a doorstep, which traditionally symbolizes a threshold and the beginning of a new status, as in the ritual of carrying a bride over the lintel. The image suggests that Afghanistan is an infant in a state of transition: It has no past yet and it stands at the beginning of its growth toward becoming a respectable nation in the global order. [21]

The representation of Afghanistan as a child also features in DoD imagery, and, as the next chapter shows, reoccurs in the press. Another neocolonial way to represent Afghanistan that occurs across these domains is feminization. Captions frequently specify the gender of depicted participants when they are women, even when there seems little reason to do so, as in "Three *female* delegates photographed during the Loya Jirga" (United States Agency for International Development, 2002b, para 1, italics added). The caption writer mentions their gender, not because it is hidden from the viewer of the image, but apparently to signal that it is important. Afghan women were central in the USAID *Out of the Shadows* exhibit that begins with a text about what it describes as "the popular image" of Afghan women:

> The most popular image was that of women cautiously lifting the veils of the once mandatory burqa to see for themselves the promise of a new Afghanistan. Although seen by the Western world as an icon of Taliban oppression, the burqa remains a common sight in Afghanistan, especially in rural areas. Yet it is now a personal choice, not a mandate. ... And although some women continue to wear the burqa, all women across the country are being given the opportunity to step out of the shadows to play vibrant and significant roles in the rebuilding of their beloved country. (United States Agency for International Development, n.d., para 2)

The opening text seeks to reframe the burqa from a symbol of oppression to one of personal choice, and, although the exhibition photographs show us women who are different in physical appearance from Western women, the text subtly suggests that Afghans should be mirror images of Americans. The exhibit claims to counter views "by the Western world" and to tell "the story of their lives today—out of the shadows—as they work to build a better life for themselves, their families, and their country" (United States Agency for International

Development, n.d., para 5) but it projects intentions on Afghan women that seem to express the individual, family, and national ethos of the United States as indicated by "for themselves, their families, and their country." Phrases like "their beloved country," which also are used, are further evidence of a nationalist ethos. "Love for one's country" may be "natural" in the patriotic culture of the United States but is probably a more alien concept to many Afghans. The *Out of the Shadows* exhibition text annihilates such differences and tends to homogenize culture, community, and place. Acculturation and assimilation are seen as key to identity formation. The text claims that Afghan women are moved "out of the shadows," but it moves them into the confines of the Western norms through which they are seen, not unlike *The Family of Man* did in the 1950s. Afghan women are made to conform to Western ideals of individual choice and to a mythical self-image of the West in which women are represented as truly free. The implied opposition seems to be one between free choice, as a characteristic of the secular modern society, and obligations as a key feature of traditional religious societies. Notice also how, at the same time, and paradoxically, Afghan women are deprived of their agency in the phrase "*being given the opportunity* to step out of the shadows" (italics added). The logic of this *empowerment* frame, one that has been criticized for being related to individualism and masculine concerns with mastery, power, and control (Riger, 2005), seems to be that Afghan women cannot have power unless it is bestowed on them from the outside.

Abu-Lughod (2002) raised the issue of Afghan assimilation in her discussion of women and the veil: "Can we only free Afghan women to be like us or might we have to recognize that even after 'liberation' from the Taliban, they might want different things than we would want for them? ... We may want justice for women, but can we accept that there might be different ideas about justice and that different women might want, or choose, different futures from what we envision as best?" (pp. 787-788). In answering this question Abu-Lughod quoted Saba Mahmood to reject the automatic acceptance of freedom and liberation as *motivational* frames: "The desire for freedom and liberation is a historically situated desire whose motivational force cannot be assumed a priori, but needs to be reconsidered in light of other desires, aspirations, and capacities that inhere in a culturally and historically located subject" (p. 788). It is this cultural and historical location that USAID photo-texts ignore when repositioning the burqa as a matter of free choice, and this stands in a somewhat paradoxical relation to the frequent use of the qualifier "local" elsewhere to describe Afghans, as in "*Local* Afghan girl photographed during the Loya Jirga" (United States Agency for International Development, 2002a, para 1), and that locks Afghan women in their past. USAID representations seem to situate the Afghan female body ambivalently at the crossroads of two myths. The first is that of globalization, the idea that an individual person can transcend time and space boundaries and become part of a Westernized global village, which the Afghan woman is requested to join. The other is that of localization, the idea that a person is entirely rooted, bounded, and local. The discourse of the USAID shifts between these two myths: It describes Afghans as "local" and wants to portray them as "global" at the same time. By reframing the burqa as a personal choice in the *Out of the Shadows* text veiling is no longer explained as a form of local tradition. Its spatial and temporal specifics are removed to make Afghan women acceptable members of a presumed global world based on originally Western economic, political, social, and cultural principles. By making the burqa a personal choice burqa-clad Afghan women are made part of the "global" despite their traditional appearance. After ungrounded expectations that Afghan women would dismiss their *chadri*, reframing the garment as a matter of personal choice allowed the USAID to present Afghan women as members of the global world. But like other dress codes, wearing a burqa is more a social practice than a matter of free individual choice. It is better understood, at least partly, as an attribute that defines the bodily self intersubjectively and intercorporeally. Such an understanding would require placing Afghan women in their proper historical context, which the USAID exhibition text generally avoids despite the frequent labeling of Afghans as "local."

## PHOTO-TEXTS IN/ABOUT WAR

In accordance with their public affairs mission, the USAID and the DoD produce, publish, and distribute images and texts that highlighted their own role. Most of the USAID photo-texts suggest a conventional non-participatory development and aid model in which Afghans figure as beneficiaries and as peripheral agents to their own development. Other USAID photo-texts focused on individual leadership and assimilation. The leadership photo-texts emphasized the role of the West; the assimilation messages suggest that Afghans are "like us" or should be.

PSYOP messages used in Afghanistan are related to DoD and USAID public affairs categories in two ways. On the one hand, some PSYOP photo-texts used in Afghanistan were construed around the theme of superiority and control. They stressed military surveillance and combat capabilities, not to instill confidence in a home audience, as in DoD public affairs, but to demoralize the enemy. On the other hand, many PSYOP photo-texts were of a nonmilitary, political, and ideological nature, produced with the intent to appease, pacify, and unite the population of Afghanistan. These messages suggested harmonious national or international relations, cooperation, and unity; they promised material prosperity[22] and liberation from oppression. These photo-texts were similar to those of USAID with a focus on the positive consequences of the intervention and to DoD photo-texts of humanitarian warfare.

The thematic convergence of PSYOP and public affairs messaging suggests a number of shared core values, beliefs, anxieties, and desires. Official U.S. representations of Afghanistan appear to be clustered around four interrelated frames: the *managerial*, *humanitarian*, *unity*, and *leadership* frames. This frame cluster suggests that war can be managed provided technologies are precise, outcomes beneficial, nations united, and leaders responsible.

First, the *managerial* frame offers a masculine vision of the world that celebrates technological might and physical prowess. DoD public affairs communication in particular stresses technological supremacy, surveillance capabilities, and physical strength. It conveys an image of an army that is ready and in control. Especially when props, such as weapons, are elevated to the status of grammatical actor, the *managerial* frame dehumanizes conflict. The emphasis on the technological in DoD and USAID communication suggests that war and nation-building are mostly technical operations, a framing that obscures their social, political, and cultural aspects. Because it suggests that conflicts can be managed without attention to these aspects, the *managerial* frame is generally a-historical. With its emphasis on the precision of weaponry and the use of proper technology, it helps to justify war by suggesting that the *jus in bello* principle of proportionality is respected.

Second, related to the *jus ad bellum* principle of just cause, the *humanitarian* frame is a positive *consequences* frame that legitimates war by drawing attention to the beneficial results of a military intervention. Humanitarian framing dominates the visual communication of the USAID but is not absent from DoD public affairs and PSYOP messaging. The frame is thematically developed in messages that emphasize its material benefits and immaterial rewards. It is realized in concrete scenes of cultural, social, and economic activities that suggest liberation, freedom, progress, and prosperity. Photo-texts with a focus on settings such as new government buildings or developmental means such as civil society infrastructure tend to "institutionalize" human development.

Third, the *humanitarian* and *managerial* frames are related to the idea of social unity, another important element in public affairs and PSYOP. The importance of social unity—at home and in Afghanistan—is suggested in photo-texts that portray individuals as family members or as members of a larger social group. Unity is presented as a precondition for properly managing war and bringing war to a positive conclusion. It is a propagandistic frame because it calls on viewers to participate in the war effort of the larger community to which they belong.

Suggesting that victories are based on collective strength, the *unity* frame homogenizes social life by establishing links between the civilian, the military, and the political order. It is a patriotic frame because it celebrates the spirit of the nation as one. The unity of the nation is used to garner support from the home front, and the family is made subservient to the greater cause of the nation. The unity frame is conveyed not only through an emphasis on national cooperation but through photo-texts that stress international cooperation as well. In both cases, national unity is approached positively. In PSYOP, the *unity* frame also is expressed negatively, through a focus on those threatening national unity.

DoD photo-texts project familial unity onto the military and the nation, and the familial model is used to portray international relations in a hierarchical way. In PSYOP, photographs of families are used similarly as models for national integration and harmonious international relations. Although USAID photo-texts are not concerned with the unity of the nation or the cohesion of the military apparatus, they do rely on the familial model to suggest transnational integration. Some USAID and DoD photo-texts invite the viewer to think about international relations as family relations whereby Afghanistan is typically presented as the child of the global family and the United States as the father.

Fourth, the *managerial*, *humanitarian*, and *unity* frames are related to a belief in individual leadership. Individual leadership and unity are presented as preconditions for properly managing war and for bringing it to a successful end. Individual elite leaders are portrayed in photo-texts of press briefings, ceremonies, or the opening of schools. They function as models within a managerial and humanitarian context, and further figure as advocates and sponsors of social cohesion. This positive image of individual leadership is reversed in PSYOP, where bin Laden and Omar are represented as evil.

Photographs of Western leaders in public affairs are an instance of domestication, and because DoD and USAID photo-texts are meant to legitimate their own organizations, domestication is generally high. The DoD and USAID frames reveal little or nothing about the ground realities in Afghanistan. In DoD and USAID photo-texts, Afghanistan appears as peripheral, appreciated almost exclusively as an exotic backdrop. Although made on location, DoD images are not about Afghanistan but about the United States. They mostly celebrate the unity and might of its forces. In USAID photo-texts, the perspective remains that of the "developer." Afghans are sometimes represented as auxiliary participants in the reconstruction process but much more often as recipients of aid and hardly ever as agents of their own destiny. Many images evoke an ethnocentric and neocolonial view of the world in which several of the traditional dichotomies are reinforced: between the West and the East, modernity and tradition, rationality and irrationality, present and past, adult and child, man and woman. Violence afflicted on Afghans or others, and its structural and cultural origins, are eclipsed. Few images depict violence, and when they do, violence is merely represented as material damage. The end result is a view of the military and humanitarian intervention as an almost bloodless and well-managed operation.

The reappearance of public affairs scenes, themes, and frames in PSYOP suggests that there are no strong boundaries between photography used *in* warfare and photography *about* war. PSYOP shares with DoD public affairs the emphasis on technological supremacy, and with USAID public affairs a focus on the material and immaterial rewards of conflict. In the first months of the intervention, audiences at home and in the war zone were targeted with similar ideological messages about the humanitarian aspects of the intervention in Afghanistan. To a considerable degree, PSYOP and public affairs messages overlapped. Aspects of the *managerial*, *humanitarian*, *unity*, and *leadership* frames occurred across PSYOP and public affairs. Moreover, as seen in the next chapter, most of these official U.S. *master* frames resurfaced in the western European press, especially the *managerial* frame. But the western European press also extended the range of official U.S. frames.

## NOTES

1. The images of the Farm Security Administration are regarded as one of the most influential collections of modern documentary photography. Originally called Resettlement Administration, the Farm Security Administration was set up in 1935 as a New Deal agency to alleviate the effects of the Depression in rural America. A historical section was created within the agency to support the agency's mission with photographic material. It was later relocated to the Department of Agriculture, and from there to the Office of War Information (see Bezner, 1999; Finnegan, 2003).
2. Originally, Edward Steichen planned an exhibition about national defense, to be called *The Arsenal of Democracy*. When the United States became involved in the war, the project was renamed *Road to Victory*.
3. For a critical discussion of *The Family of Man*, see Barthes (1973), Sekula (1984), and Bezner (1999), who described the response to the exhibition as "forty years of hostility" (p. 173).
4. The exhibition consisted of 28 images. The World Trade Center Archive includes more than 8,000 pictures.
5. Signed by President Harry S. Truman, the Smith-Mundt Act is officially known as the U.S. Information and Educational Exchange Act U.S. of 1948 (Public Law 402, amended in 1972 and 1998).

6. The Coalition Information Centers were turned into the now also defunct Office of Global Communications in July 2002.
7. Committees included the Strategic Communication Policy Coordinating Committee, the National Security Council Policy Coordinating Committee, the Muslim Outreach Policy Coordinating Committee, and the Interagency Strategic Communication Fusion Team.
8. The "war on terror" discourse was used consistently by Bush (Domke, 2004) but questioned by various officials as the wars in Afghanistan and Iraq continued (Stevenson, 2005). Discussions about proper terminology for the intervention in Afghanistan and its associated activities found their way in documents of the Department of Homeland Security (2008) and the National Counterterrorism Center (2008). Finally, newly elected President Barack Obama ended the use of the "war on terror" in early 2009 (Priest, 2009).
9. The just-war literature varies on what is an acceptable "just cause" and what is not. Narrow interpretations may limit "just cause" to national self-defense, whereas broad interpretations may accept situations of extreme oppression (Rengger, 2002). The relation between *jus in bello* and the doctrine of double effect is the subject of debate as well. See Ceulemans (2007). The modern classic of just-war theory is Walzer (1977).
10. The Patriot Act is officially known as Uniting and strengthening America by providing appropriate tools required to intercept and obstruct terrorism (USA Patriot act) Act of 2001 (Public Law No. 107-56).
11. See the Homeland Security Presidential Directive-3, March 2002, and the Homeland Security Presidential Directive/HSPD-5.
12. Photographer Nina Berman (2008) explored these politics of fear in *Homeland*. See also *Fear This* by photographer Anthony Suau (2004).
13. The idea of limited war was first voiced by the British military historian Liddell Hart (1895-1970), who believed that nuclear weapons had made total war obsolete (Bond, 1977). The United States further developed the concept as a military strategy for the Cold War. Limited wars are supposed to have limited objectives, limited impact on civilians, and they do not draw the entire population into the war effort. They are bounded geographically and require a limited amount of military resources. The controversy surrounding limited wars, then, typically concerns its boundaries.
14. *Actor* is a term from systemic functional linguistics used to designate the originator of a material process in representations (Halliday, 2004). In systemic functional linguistics, the term is usually written with a capital "A," like other grammatical categories.
15. I use the term *fetishism* here to indicate the tendency to ascribe power to technological devices and the belief that these powers can be used to control others.
16. These terms are borrowed from functional grammar and designate those affected by a process (Halliday, 2004).

17. Information Operations grew out of the Information Warfare and Command and Control Warfare concepts of the late 1970s and are now regarded as one of three activities to achieve "information superiority," alongside Information Management and Intelligence, Surveillance, and Reconnaissance (United States Army War College, 2006).
18. A photograph of a child holding up this leaflet was published in *Afghanistan: The Road to Kabul* (Haviv, 2002a).
19. The photograph of the Muslim women with the candle also was published in *De Morgen*, with a caption that criticized the television networks: "Half the world is on fire, but first the power center in New York has to be destroyed before television networks get into action" (September 21, 2001).
20. The USAID/Afghanistan Strategic Plan 2005-2010 (United States Agency for International Development, 2005) mentioned three core strategic objectives of the mission: a thriving economy led by private business, a democratic system, and improved education and health care. Since 2001, the USAID has tried to implement this tripartite strategy through grants and contracts to NGOs and contractors, a process that has turned out to be hazardous.
21. In the United States, the word *reborn* might evoke connotations associated with the fundamentalist Christian notion of redemption. It might suggest that the Afghans are forgiven for their misbehavior and receive a new chance.
22. *Material prosperity* also was a theme in the public diplomacy Shared Values campaign, launched in November 2001 by then Undersecretary of State for Public Diplomacy Charlotte Beers. The campaign's television ads portrayed American Muslims and were targeted at the world's Muslim population. Produced by the advertising agency McCann-Erickson, they were launched in Indonesia at the time of the Ramadan, and also ran in Pakistan, Malaysia, Indonesia, and Kuwait. Al-Jazeera and other Arab networks refused to broadcast it. The campaign eventually was suspended. Other public diplomacy initiatives to promote American values included the youth-oriented Arabic-language magazine *Hi*, the student exchange program Partnerships for Learning, Radio Sawa, and Alhurra TV.

3

# Western European Press Views

In the previous chapter, I argued that the discursive borders between public affairs and PSYOP are porous. This chapter shows that the boundaries between public affairs and PSYOP, on the one hand, and photojournalism, on the other hand, are under pressure as well. To some extent, the press framed the conflict in Afghanistan in ways that were similar to those in public affairs and PSYOP. This chapter traces the similarities and differences between these areas and across eight western European elite newspapers by examining the portrayal of war leadership, combatants, and civilians.

## PHOTOGRAPHY AND THE NEWS

The photo-texts discussed in this chapter were taken from the print editions of the British newspapers the *Times* and the *Guardian*, the French papers *Le Monde* and *Le Figaro*, the German papers *Die Welt* and *Die Zeit*, and the Belgian papers *De Standaard* and *De Morgen*. One half of these newspapers is generally regarded as left or left centric (*Guardian*, *Le Monde*, *Die Zeit*, *De Morgen*), the other half as more conservative (*Times*, *Le Figaro*, *Die Welt*, *De Standaard*). The newspapers are all known as important shapers of public opinion in their countries, and they are usually seen as influential in forming the political agenda of the national establishment. Despite their status of national newspaper, most of these publications do not have any straightforward relation to national culture. On the basis of their national print circulation and readership figures,[1] *Le Monde* and *Le Figaro* probably are better characterized as Parisian than French, the *Times* and the *Guardian* are more English than British (Rosie, MacInnes, Petersoo, Condor, & Kennedy, 2004), and *De Standaard* and *De Morgen* more Flemish than Belgian.[2] The German newspapers *Die Zeit* and *Die Woche* have a more obvious national profile than the others but their position is challenged by a number of important political weeklies, such as *Der Spiegel* and *Stern*, and by a strong regional press.

Historically, the German news model is related to the Anglo-Saxon model. After World War II, British and U.S. press officers trained German journalists and emphasized the need to keep opinion and fact separate. According to Esser (1998), this Anglo-Saxon influence has been mediated by the traditional German organizational structure of the newsroom and, more generally, by German philosophical traditions and experiences that favor a plurality of worldviews. Esser concluded that "the divorce between editorial opinion and factual reportage in German newspapers is artificial, just external, because the writer of both accounts is often the same" (p. 398), and that personal preferences of a *redakteur* (*editor*) are more likely to enter the news than in the case of the British press, where newsrooms are organized according to a stronger division of labor for writing, subediting, revision, and checking. Although the entire European newspaper market has been described as "a bloody battleground" (Levine, 2009, p. 27), the British press in particular is characterized by fierce market competition and a clear divide between elite papers and popular tabloid papers, even though differences appear to have decreased (Franklin, 1997). Competition and centralization are regarded as major reasons for its sensationalism (Esser, 1999).

The other countries also have their own distinct newspaper traditions. The French tradition has been shaped by relatively strong government interference. Between 1970 and 1990, total state aid amounted to 10% to 15% of total press revenues, one of the highest levels of support in Europe (Benson, 2002). Benson identified the dominant literary and scholarly circles in Paris as another major determinant of the French press tradition, and he saw their influence reflected in the use of narrative formats and a mixture of descriptive and normative statements. In Belgium, language, philosophical, and political divisions have largely determined the press landscape. The roots of *De Standaard* are Flemish-nationalist and Catholic. The newspaper was first published in December 1918, after an earlier attempt to establish a Catholic newspaper was put on halt because of World War I. *De*

*Morgen* started more than half a century later, in December 1978, under the impulse of socialist leaders in Flanders. Its first editor-in-chief was 1968 student leader Paul Goossens, who followed a party-independent line and used *De Morgen* to promote progressive forces in Flanders.

As this brief outline indicates, whereas European newspaper traditions differ markedly from each other, partisanship and a more analytic mode of reporting have been common to all. In the United States, news journalists generally have strived to adhere to an ideal that keeps information and commentary separate (Josephi, 2005). An analysis of a random sample of political news stories in *Le Figaro* during 1995, 1996, and 1997 demonstrated that about one-third of the paragraphs offered interpretation or opinion against 10% in a comparable sample of news stories from *The New York Times* (Benson, 2002). The American objective model for news journalism often is seen as the hegemonic one, but perhaps print newspapers will be forced to move toward a more analytical, European approach due to increased competition in hard news from online sources. This could affect the role of photography in print. It might stimulate a trend toward more analytic, symbolic, or rhetorical imagery, although currently there seems to be little evidence that this is happening. On the contrary, complex visual narratives generally are avoided in popular and elite news. Few print publications today—not just newspapers—seem willing to invest in staff reporters and complex forms of reporting. The photography of conflict is particularly expensive, which means there is usually little or no room for experimenting. As a result, conflict photography remains a largely conservative area, dominated by tradition and commerce, and driven by a desire to capture action, particularly when it has spectacular material and physical consequences.

To some extent, the visual culture of elite papers remains grounded in a distrust of images. Karin Becker (2003) argued that, in the context of the elite press, "photography's more immediate, direct appeal is seen as a threat to reason, and to the journalistic institution's Enlightenment heritage" (p. 291). She called this position paradoxical because photographs have been valued at the same time for their evidentiary quality and objectivity. A number of techniques, such as the use of rectangular pictures in isolation, are used in the elite press to "control" the image. This level of control is absent in popular newspapers, which have integrated visuals enthusiastically in their layouts. Headlines and graphic elements are laid on top of the pictures, and various forms of manipulation such as cropping, cutting, and montage are used to attract attention. Although the elite press does not routinely use these techniques, they do consider the visual indispensable today. If perhaps only from a marketing point of view, for all elite newspapers, photography, illustrations, and design are important factors. The market is one of several external factors that determine the use of photography in the news.

Surrounded by suspicion at least since Plato's allegory of the cave, visual images have always been subjected to control and censorship. Especially since the Vietnam War, political and military leaders tend to assume that press coverage influences policies and operations. They generally seem to believe that there is a strong correlation between political and military success on the one hand, and reports of casualties on the other hand. Media management strategies therefore are often focused on averting responsibility when casualties occur and on bringing the humanitarian dimension of warfare under attention (Shaw, 2002). Control of the media traditionally has been exercised through direct censorship and accreditation (Hallin, 1997), a dual system that began to take shape during the American Civil War and was further developed during World War I (Hallin, 1989). In World War II, photojournalists were accredited to the armed forces and to official agencies such as the British Ministry of Information and the U.S. Office of War Information. Because of the widespread use of control mechanisms, George Roeder (1993) called World War II "the censored war" in contrast to "the uncensored war" (Hallin, 1989) in Vietnam. During the Vietnam War, reporters had to be accredited if they wanted to accompany soldiers, but they were not subjected to censorship. Photographers could move around quickly and compete with television makers with their more complex equipment. International photo agencies such as Magnum Photos and Black Star rose to prominence but the Vietnam War eventually also coincided with the decline of the picture magazine and the photo-essay. Although President Nixon held photographers and cameramen responsible for dividing the public and for the eventual retreat of the United States from Vietnam, several studies later indicated that the media were actually mostly supportive of the Vietnam War and became critical only when elite consensus had already broken down (e.g., Braestrup, 1983; Hallin, 1989). Even My Lai television coverage was "usually cautious and dispassionate" (Hallin, 1989, p. 180).

Despite evidence to the contrary, the idea that the media can make or break a war has continued to determine relations among the government, the military, and the media.[3] After Vietnam, the Falklands/Malvinas War (1982) and the invasions of Grenada (1983) and Panama (1989) marked a new era of stricter political, military, and corporate control over photography, often legitimated for security and protection reasons (S.D. Cooper, 2003). During the Falklands/Malvinas War, the British accredited only two photographers, excluding the renowned Don McCullin (Morrison & Tumber, 1988). Only 202 photographs were transmitted from the war zone, none were made of the initial part of the conflict (S.D. Cooper, 2003; Harris, 1983), and the

British press did not publish photographs of British casualties (Taylor, 1991). Grenada and Panama repeated this pattern of control. During the invasion of Grenada, on October 25, 1983, military officials denied photographers access to the island. Two days later, a small number of reporters were allowed to enter, but it was not until November 7, 1983, when the fighting was over, that unrestricted access was made possible. This led to protest among journalists and eventually to negotiations with U.S. military officials in the Sidle Panel,[4] which laid the foundations of the DoD National Media Pool. The pool was supposed to help journalists cover the initial stages of a conflict, but it failed in Panama and became operational there only after the initial combat was over. During the Persian Gulf War (1990-1991), media pools were expanded to 1,500 journalists, but access to people and places remained severely limited (Fialka, 1991). Hallin (1997) found that most studies of Gulf War coverage characterized it as "sanitized, heavily dominated by official sources and perspectives, and highly favorable to the military operation" (p. 215).

To answer complaints about the press pool and the exclusion of the media from the battleground, the DoD introduced embedding after the Gulf War. An embedded reporter is assigned to a military unit and stays with it for some time, usually between 10 and 30 days. The DoD saw the new approach as a way to counter adversarial propaganda and create support for its activities. In 2002, the military and the press agreed on the embedding strategy developed by Assistant Secretary of Defense for Public Affairs, Victoria Clarke, and her Deputy Secretary, Bryan Whitman. A form of embedding had been introduced earlier, during the 1996 U.S. military operations in Bosnia (Paul & Kim, 2004), and further experiments had taken place in Kosovo in 1999. In Afghanistan, the DoD embedded a small group of reporters during Operation Anaconda, the battle against Al-Qaeda and Taliban members in the Shashi Kot valley near Gardez in March 2002. Although several critics argued that embedding was against the American First Amendment (Zeide, 2005), the system was fully implemented in 2003 during Operation Iraqi Freedom. Different national forces have since then set up their own embed policies with different forms of accreditation, supervision, and censorship (The Hague Centre for Strategic Studies, 2008).

Reviewing embedded reports, Fox and Park (2006) suggested "standards of impartial inquiry appear to have been maintained by the embedded reporters, even when the technique of detached writing style was not used" (p. 48). Other research found that embedded reporting is significantly more positive toward the military (Haigh et al., 2006). This was explained elsewhere as a function of the closer interpersonal relations between journalists and military personnel (Pfau et al. 2008). Pfau et al. found that embedded coverage is more favorable in tone toward the military generally as well as in its depiction of individual troops. They also argued that embedded journalists show less interest in broader aspects of a conflict, in its purpose, and in its impact, resulting in more decontextualized coverage and *episodic* framing. Although more journalists may have gained access to troops and operations through embedding, the scope of their work has narrowed, according to a study on press coverage of Afghanistan in Dutch newspapers (The Hague Centre for Strategic Studies, 2008). The study found that 90% of all embedded reports focused on the military and the military mission, and 80% of their sources were military officials.

The impact of embedding extends beyond the tone and content of reports. Close interpersonal contact occasionally led to some embeds joining the fight, and nonembedded or *unilateral* reporters found it more difficult to work.[5] Critical embedded reporters were sometimes prohibited from rejoining troops later. Britain's Ministry of Defense barred news broadcaster ITN from embed programs in Afghanistan and Iraq after its ITV News channel refused to retract critical coverage of the care for wounded soldiers in a series called "Betrayed? An Investigation" (Knight, 2006). Another problem is that embedding may make the press more dependent on the military because embedded journalists do not develop their own know-how and networks to operate independently in conflict zones. The overemphasis on the military in embedded reporting may also encourage an overestimation of the role of the military in solving conflicts, which could encourage military spending.

The military do not merely exercise influence via embedding. They exercise direct influence on media content through the distribution of their own military photographs. Their publication in newspapers is now a common practice, and press captions rarely seem to differ substantially from those attributed to the pictures by the DoD.

The photograph of two soldiers running down a street in Fig. 3.1 is an example. The DoD captioned this image as "U.S. Marines raid a suspected Al Qaeda hideout in the Helmand Province of Afghanistan," keeping the focus on the action of the moment captured by the photographer. The reader is not offered any contextual explanation that might clarify the purposes and consequences of the action. The reader is not informed about what happened before or after the raid, and why it occurred. Agence France-Presse distributed the picture, and *Le Figaro* published it with a caption that described it as "American marines are conducting since Monday evening an intelligence mission targeted at a camp presumably serving Taliban and members of the Al-Qaeda network", adding some minor details but no substantial contextual explanation.

Sometimes military images were subtly reframed. One DoD picture of soldiers preparing an aircraft for a strike mission (Fig. 3.2) was republished with some cropping on *Die Welt*'s front cover. Although the DoD's original caption focused on the precision of the bombing, the newspaper's caption focused on its scale: "*New wave* of attack: The

**Fig. 3.1. "U.S. Marines raid a suspected Al-Qaeda hideout in the Helmand Province of Afghanistan." Johnny Bivera (2002)**

DoD image published in *Le Figaro* on January 5-6, 2002 (p. 3, "The interminable hunting of Mullah Omar")

**Fig. 3.2. "Aviation ordnancemen on the USS Enterprise." Douglas Houser/U.S. Navy (2001)**

DoD image published (in a cropped version) on the front page of *Die Welt* on October 10, 2001

crew of the aircraft carrier USS Enterprise prepares a fighter plane for *further deployment* in Afghanistan. Pilots are slowly *running out of targets*" (italics added).

The use of military photographs is an inexpensive way for newspaper publishers to visualize a conflict. But military photographs are valued for more than economic and pragmatic reasons. Military images also are appreciated for their aesthetic qualities. A series of military photographs that focused on soldiers and their military equipment—not part of my corpus—succeeded in winning third prize in a contest for photojournalists in The Netherlands in 2007.[6] The nomination of a military photographer in a contest for photojournalists illustrates the pervasiveness of military imagery. This militarization of journalism can be seen as part of a larger discursive militarization (Lutz, 2002) that in its mass media form took off after Vietnam, when war became a major theme in the entertainment industry, and entertainment became a site of strategic importance for the military. Although militarization of the cultural industries does not seem to have reached the same proportions in western Europe as it has in the United States, the western European press seems to rely increasingly on military images and sources to tell its stories too, and has employed the same or similar frames as the military.

Journalists who wanted to cover the invasion of Afghanistan in 2001 faced many difficulties, including security problems, language and cultural barriers, commercial demands, and financial constraints. On the one hand, the Taliban regime imposed severe restrictions and did not allow any foreign journalist into the country from September 2001 to October 13, 2001, when they invited the press for a brief tour to Koram, a village in eastern Afghanistan that was reportedly bombed by the United States. On the other hand, the U.S. government and military denied journalists access to troops and conflict zones until late November 2001. The severe restrictions had a defining impact on mass media content, especially in the first weeks of the war. Journalists were stuck at the Afghan border, which explains the relatively high number of photographs of the refugee camps in Pakistan and of demonstrations in the streets of Islamabad, Quetta, and other cities in Pakistan. Although journalists covered life in the Afghan refugee camps, they faced access restrictions there, too. When allowed to enter, an armed government security officer accompanied them. Officials argued that the restriction was a protective measure because many of the refugees were angered by the allied bombardments and because some journalists had already been attacked.

The initial stages of the intervention in Afghanistan remained largely invisible, as in the case of the Falklands/Malvinas War, the Panama and Grenada invasions, and the Persian Gulf War. According to Neil Hickey (2002), journalists were denied access to U.S. troops to a degree previously unseen. The Committee to Protect Journalists (2002) called access to information about U.S. military operations "even tighter than during the tightly scripted Persian Gulf War'" (para 13), and it argued that U.S. Defense Secretary Donald Rumsfeld and other officials invoked the special nature of the "war on terror" to legitimate restraints on information. Victoria Clarke defended the approach arguing that what could be covered was covered extensively (Hickey, 2002). Before and at the beginning of the war, the United States allowed some reporters on their ships where missiles were launched, and some journalists took part in Airborne Warning and Control System flights over Afghanistan. Access to bases in Pakistan, Uzbekistan, and other countries was denied because, according to the Pentagon, leaders there worried that news reports could cause political unrest (The Associated Press, 2001b). Generally, journalists did not have proper access to land and sea bases from which air attacks were launched, and they could not interview pilots and members of Special Operation Forces, whether located in Pakistan, Tajikistan, and Oman (Hickey, 2002). Two-thirds (68.7%) of the respondents to an online survey among 516 members of the National Press Photographers Association agreed that photojournalists did not have the necessary access to photograph the war in Afghanistan (Fahmy, 2005). The written press received most of its information from press briefings in Doha (Qatar) and the Pentagon.

The nature of the operation also had a considerable impact on press coverage. The initial intervention in Afghanistan was not a frontline battle but a large-scale bombing campaign. This type of warfare, combined with the denial of access to troops, made photography virtually impossible. On the one hand, the initial actions on the ground by Special Operation Forces were declared sensitive and kept secretive. On the other hand, it is impossible or dangerous to be too close to the targets of a large-scale air campaign, and access to carriers and airstrips usually was denied for safety reasons. To some extent, the initial situation in Afghanistan was similar to the Kosovo War of 1999. But whereas the military failed to control commercial satellites during the Kosovo War, they succeeded in keeping images taken from the sky out of the press during the intervention in Afghanistan by buying the exclusive rights.

The Taliban also made covering the conflict difficult. When the intervention in Afghanistan started in 2001, they already had a firm reputation for being hostile toward journalists, particularly photographers and cameramen. Many Afghan journalists went into exile during the Taliban rule, and according to Reporters sans frontières (Reporters Without Borders, 2001), the regime arrested more than 25 foreign journalists between September 1996 and February 2001. When allowed inside the country, foreign journalists were forced to work with officially appointed translators. They had to stay in Kabul's

Intercontinental Hotel, and could not film or photograph any people, or talk to women without permission. At the time of the invasion, there were no more television stations in Taliban-controlled Afghanistan. The Taliban had banned the Internet too, even though they had its own official Web site at afghan-ie.com until shortly after 9/11. Only the government radio station, the Voice of the Sharia, and a few print publications were in operation.[8] Most of the newspapers from the Soviet time had already disappeared by the time the Taliban came to power. The most important Taliban medium, the Voice of the Sharia, located at the broadcasting center of the former Soviet Radio Afghanistan, was one of the first U.S. targets on October 8, 2001. After its destruction, the Taliban continued to spread its message through the Afghan Islam Press agency in Pakistan. The Taliban also continued to broadcast from Mazar-e Sharif until the Northern Alliance troops took over the city on November 12, 2001.

Just before 9/11, a relatively high number of foreign journalists were in Afghanistan. Two events had attracted their attention: a trial of eight aid workers charged with Catholic proselytizing, and the killing of Northern Alliance military commander Ahmed Shah Masud by suicide bombers pretending to be journalists. When accused of protecting bin Laden and of collaboration with Al-Qaeda, the Taliban regime asked all foreigners, including journalists, to leave the country. Several journalists who sneaked into the country nevertheless were arrested and charged with espionage. On October 13, 2001, the Taliban changed its press policy and invited a group of foreign journalists to Koram, which was reportedly damaged by coalition air strikes. The journalists also were taken under the supervision of the Taliban government to the city of Jalalabad. One month later, on November 20, 2001, approximately 100 foreign journalists were allowed to enter Spin Boldak, a town and refugee camp in eastern Afghanistan near the border with Pakistan. Two days later, they were escorted back to the Pakistani border town of Chaman, and a scheduled visit to Kandahar was cancelled.

Approximately 700 journalists had entered Pakistan before the first air strikes on October 7, 2001 (Menon, 2001). CNN, BBC, Fox News, and other television networks reported from the rooftop of the Marriott Hotel in Islamabad. Some journalists later left their hotel because they feared it would become a target. Other journalists flew into Tajikistan and tried to arrange transport to get into the Northern Alliance-controlled territory of Afghanistan. Only a few local journalists were inside Taliban-controlled Afghanistan. The Associated Press, Reuters, and Agence France-Presse had retained their offices in Kabul during the Taliban regime, and their Afghan staff was able to continue reporting. Among them were Amir Shah of the Associated Press, Sayed Salahuddin of Reuters, and Said Muhammad Azam of Agence France-Presse. The Arabic-language Al-Jazeera also stayed on in Kabul and was able to play a central role in the Afghan media war, challenging the position of established Western news organizations. Founded in 1996, the network quickly became known with its broadcasts of videotapes from bin Laden and Al-Qaeda, which led to accusations of propaganda by the Bush administration. On November 13, 2001, the day Kabul fell to the Northern Alliance, the United States bombed the Al-Jazeera office in Kabul and damaged the news bureaus of the Associated Press and the BBC. According to Ron Suskind (2006), the strike was conducted on purpose but the United States denied the allegation. Gen. Richard B. Myers, then chairman of the Joint Chiefs of Staff, explained in a letter to the Committee to Protect Journalists (2006) in February 2002 that the military believed the building to be an Al-Qaeda facility (para 7). Suspicion that the United States had deliberately targeted the Arabic broadcaster rose when the United States later hit the Al-Jazeera headquarters in Baghdad on April 8, 2003, as well as another Arab network station, Abud Dhabi. Suspicions were fueled further when the *British Daily Mirror* (Maguire & Lines, 2005) reported an April 2004 conversation between President Bush and British Prime Minister Blair in which the former suggested bombing Al-Jazeera's office in Doha, Qatar. But a senior U.S. diplomat in *The Washington Post* dismissed the charge, saying that it sounded "like one of the president's one-liners that is meant as a joke" (K. Sullivan & Pincus, 2005, para 3).

Because the vast majority of photographers were outside the war zone, the media had very few photographs to publish. Newspapers resorted to portraits of world leaders, photographs of military training, and pictures of demonstrations across the globe. Newspapers published photographs delivered by military officials, used video frames grabbed from TV broadcasts, and published older pictures, sometimes from a different war. For example, *Die Welt* published a photograph of military operations in Kuwait to accompany an article entitled "Have U.S. troops landed yet?" (October 19, 2001). A number of photographs, accompanying stories about Afghanistan, were from the period of the Soviet invasion. British newspapers also used images from the Great Game. Newspapers around the world could neither show much of the Western troops or their combat activities nor Afghan life during the wide-scale bombing campaign. Pictures of Northern Alliance soldiers were published more often. The Northern Alliance troops, which at the beginning of the invasion controlled about 5% to 10% of the territory, gave access to Western journalists and transported several of them from Dushanbe, the capital of Tajikistan, into Afghanistan.[9]

Journalists were allowed to accompany U.S. soldiers for the first time inside Afghanistan on November 25, 2001, to Camp Rhino, near Kandahar. The pool consisted of reporters from the Associated Press, the (Baltimore) *Sun*, CBS, CNN, *Newsweek*, *The New York Times*, Reuters, *The Wall Street Journal*, Gannett, *The Washington Post* and the Associated Press Television News. The pool agreement stipulated that the journalists could not report in ways that would jeopardize the security of the base and its soldiers. Even though reporting on casualties was not forbidden, the pool journalists were nevertheless prevented from reporting a "friendly fire" incident with wounded U.S. and Afghan soldiers on December 5, 2001. The International Press Institute, Reporters sans frontières, and the *American Journalism Review* reported similar incidents that complicated reporting in Afghanistan.[10]

Interest in Afghanistan began to decrease as early as 2002 and this decline accelerated when the war in Iraq began in 2003. Although the number of casualties in Afghanistan steadily grew, Western media decided to follow the political agenda and shifted their attention to Iraq.

## INITIAL RESPONSES ACROSS THE ATLANTIC

The Bush administration took various steps to influence media coverage from the very beginning of the intervention. The press reported that the U.S. State Department tried to stop the airing of an interview with Taliban leader Omar by the government-funded radio station Voice of America, that Secretary of State Colin Powell tried to influence Al-Jazeera's broadcasting when he visited Sheik Hamad bin Khalifa Al Thani, emir of Qatar, and that White House spokesman Ari Fleischer publicly condemned ABC talk show host Bill Maher for comparing the 9/11 attacks to the U.S. strikes on Afghanistan. Commenting on President Bush's description of the hijackers as "cowards," Maher had said that launching missiles from 2,000 miles away was cowardly. Maher had to apologize, many TV stations suspended the show, and sponsors like Federal Express and Sears withdrew. Finally, his program was taken off the air (Paulson, 2001).

Media watch group FAIR qualified the attempts to influence the media in the United States as "broad" (Hart & Ackerman, 2001), and the efforts to influence apparently went beyond the military operation. Journalists covering the two other major stories of that time, the aftermath of 9/11 and the anthrax bio-terrorism threat,[11] found it difficult to do their job. In her assessment of the coverage of the anthrax crisis, Patricia Thomas (2003) reported tight government control of health and science news. She feared that the way the government handled the anthrax crisis was part of a bigger scheme and that, once the "war on terror" was over, access to government scientific research would remain limited. Although several U.S. media had taken legal steps against government restrictions during the Persian Gulf War, they now often practiced self-censorship (Knightley, 2002) and took over the Bush administration's rhetoric. Larry Flynt, publisher of *Hustler* magazine, and Getty Images did go to court to challenge press restrictions during the intervention in Afghanistan, but without success.[12]

In the wake of the 9/11 attacks, most media saw it as their duty to protect national security. According to *The Washington Post*, CNN Chair Walter Isaacson ordered his staff to balance images of civilian casualties with reminders that the Taliban had harbored terrorists (Kurtz, 2001). And at a conference in Singapore, Rena Golden, executive vice president and general manager of CNN International, admitted that there was a reluctance to criticize anything (Tomlin, 2002). Fox TV News (2001) questioned the relevancy of civilian casualties in conflict reporting, and other journalists openly declared their sympathy for the government and the war. CBS Evening News anchor and managing editor Dan Rather did so on the *Late Show* (September 17, 2001) and on *Entertainment Tonight* (October 2, 2001). *The Washington Post* withheld information from a classified briefing about possible new terrorist strikes, and newspaper and Internet publisher Knight Ridder delayed publication of a story about Special Operations units that had secretly entered Afghanistan (The Associated Press, 2001a). Such voluntary cooperation by the press is not exceptional when a war is seen as "just" and "total." All members of society are then expected to participate, including photojournalists.

Although western European regimes generally appeared to be in agreement with the military intervention by the United States in Afghanistan, the media responses on both sides of the Atlantic were different. In the first days after the attacks on the twin towers in New York and the Pentagon in Washington, the western European newspapers expressed solidarity and sympathy with the United States. In France, on September 12, 2001, *Le Monde* headlined "We are all Americans," and the newspaper reminded its readers of the declaration of solidarity expressed by President John Kennedy: "We are all New Yorkers, as surely as John Kennedy declared himself, in 1963 in Berlin, to be a Berliner" (Colombani, 2001). Quoting social democratic politician Peter Struck, *Die Welt* repeated "We are all Americans" (Dausend & Haselberger, 2001) in its September 13, 2001 issue above a photograph of members of the German government standing together in silence to commemorate the victims of the 9/11 attacks. The text below stressed unanimous solidarity with the victims of the attacks: "All fractions condemn the attacks" and "In an emotional special assembly of parliament, representatives of all fractions call for solidarity with the victims." Displaying transatlantic friendship, *Le Figaro* published a photograph of French President Jacques Chirac and U.S. President Bush on its front page of September 20, 2001, concluding in the headline that Chirac was "whole-heartedly with America." The caption, however, did not frame the attacks as war, as the Bush administration did, but pointed to Chirac's interpretation of them as "crimes against humanity."

The coverage of the *Times* was more war-oriented, especially in the first weeks after 9/11. On September 13, 2001, the newspaper was already discussing "the military options for President Bush," while *Le Figaro* waited until the end of the month to deal with "the possible scenarios" for war (September 29, 2001). Suggesting total war and emphasizing the need for leadership in times of war, the September 17 issue of the *Times* had a close-up of Blair on the cover, along with a headline with an all-inclusive "we": "We are at war with terrorism." The subheads on the front page were in the same war-minded spirit: "Blair vows to hunt down the murderers of 5,000 people"[13] and "Taliban have only days to hand over fanatic bin Laden." The lead of the cover story started with "the countdown towards America's fight back against terror begins today." On September 18, the *Times* took over Bush's words, "WANTED: DEAD OR ALIVE," and put them in block letters on its front page, leaving out the quotation marks. The *Times* accepted the war frame and even seemed to welcome it. It used the word "war" frequently throughout the month of September even though the official British declaration of war came only on October 1, 2001. War headlines and subheads included: "Terror in America: ready for war" (September 16, 2001), "Blair's finger on the trigger," "Prime Minister: No option but to act" (September 21, 2001), "America at war: Military gets ready," and "Shock troops prepare for the descent to battle" (September 22, 2001). Although some of these headlines, such as "Blair's finger on the trigger" and "Prime Minister: No option but to act," emphasized Britain's role, the newspaper framed the conflict regularly as a war of the United States, and in one caption narrowed it further down to "President Bush's war."

The *Guardian*'s response was much more restrained than that of the *Times*. While the *Times* headlined "When war came to America" after the 9/11 attacks, the *Guardian* captioned the twin towers on fire in a minimalist way as "Manhattan, 2001." In the same issue, preferring "attack" rather than "war," the *Guardian* reported: "U.S. rallies the west for attack on Afghanistan." On September 20, 2001, it concluded that the "last path to peace" was closed. Peace seemed never an alternative for the *Times*. In 2001, after 9/11, the *Times* published one picture of a peace protester, an old woman carried away by two British policemen. The image was captioned "An anti-war protester is removed from the Welsh Assembly before Mr Blair's arrival" (October 31, 2001). In the same period, the *Guardian* published eight photographs of peace demonstrations in the West. Other newspapers had more peace demonstration pictures too. *De Standaard* published five pictures of protests and an article entitled "Peace demonstrators kept dark" (October 13-14, 2001) about the absence of demonstration reports in U.S. media. Some experts argued in the newspaper that retaliation was too simple an answer (Tom Sauer, September 14,

2001). *De Standaard* also noted that even New Yorkers did not like Bush's war rhetoric and were "horrified" (September 19, 2001).

In Belgium, *De Morgen* responded with reservations about the prospect of war too. It headlined its September 12 issue "Apocalypse Now" and used a question mark "At war?"

Skepticism about the Bush administration rapidly grew, and sympathy with the United States turned out to be short-lived. In *Le Monde*, Gérhard Chaliand questioned the war frame before the intervention had begun (September 18, 2001), and Marie-José Mondzain contested the earlier proposition that "We are all Americans" by stating: "I do not feel American" (September 19, 2001). The western European newspapers expressed doubts about the military operation and the humanitarian intervention. In the *Times*, Peter Riddell looked behind the image of "the prime minister's self-confidence and authority as a war leader" and concluded that "A brave face cannot hide the enormous risks we face" (October 9, 2001). The *Guardian* published the first photograph of a victim of the war, and reported that 20 civilians might have been killed. On October 10, 2001, *Le Monde* headlined that NGOs were "in disagreement with the humanitarian war," and the *Times* reported: "Bishop questions allied bombing campaign" (October 12, 2001). The *Guardian* concluded that the "week of bombing leaves us further from peace, but no nearer to victory" (October 15, 2001). On October 16, 2001, *Le Monde* headlined that the U.S. strategy was "contested." A cartoon by Pancho on *Le Monde*'s front page depicted a blind Uncle Sam walking with a bomb instead of a guide dog, shouting "Come on, find bin Laden." On October 22, 2001, *Le Figaro* spoke of a "humanitarian drama behind closed doors" in an article accompanied by a photograph of a young Afghan refugee being chased by Pakistani border police. *Le Monde* had earlier headlined "Afghanistan: humanitarian drama" on its front page (October 3, 2001). On October 25, *Le Monde* concluded that there was a worldwide "unanimous condemnation of the military operation against Afghanistan." The newspaper had spoken out against the use of cluster bombs a day earlier calling it "morally odious and extremely counterproductive." The *Times* headlined "Smarts bombs can be stupid" and "War descends like a scythe on innocent Afghan village." On October 29, 2001, it ran the story of "An errant U.S. bomb, a stopped clock, and a funeral instead of a wedding" accompanied by a photograph of Afghan men carrying a coffin with the body of a young Afghan woman, Koko Gul, aged 22. The caption allowed the deceased's husband to speak directly to the reader: "Why cannot the Americans be more careful?" Although the caption was critical of the U.S. war campaign, it criticized only its precision. Similar critique on the nature of the bombing campaign was found throughout the western European press. On October 30, 2001, the *Guardian* concluded: "Majority want bombing pause."

The bombing continued relentlessly, however, and as evidence of civilian casualties could no longer be ignored, the press became ever more critical of the military strategy. After 3 weeks of heavy bombing, there were few successes to report. The Taliban had not been removed from power, anthrax was threatening the world, bin Laden had not been captured, and a Red Cross compound had been bombed twice. Foreign support for the U.S.-led operation seemed in decline. The war intensified the following month, and this led to more casualties, more debate among the political elite, and more critical reporting. The *Times* reported that Blair's war effort was now "under friendly fire" (November 2, 2001), and the *Guardian* quoted French President Jacques Chirac who had said that military action was "not the only way to fight terrorism" (November 6, 2001). The fall of Kabul did not change the mood. The entry of the Northern Alliance into the capital was on the front page of all newspapers. All had reservations about the city's "liberation." Captions framed the war in terms of defeat as well as victory. The *Guardian* described the entry into Kabul ambiguously as "Freedom, joy—and fear" (November 14, 2001). The *Times* took a similar ambiguous position and called it "Kabul's nervous celebration" (November 14, 2001). One headline framed the "liberation of Kabul" as in-between horror and pleasure: "Death and delight on road into Kabul." The photo caption was more unambiguously positive, focusing on the celebration of Kabul residents and the warm welcome that the *mujahedin* enjoyed. But a streamer expressed doubts: "At least the Taliban didn't burn down your business. If the Alliance control their soldiers, I will remember this day with joy. But it's too early to say if they have changed from last time." *Le Figaro*, which used a popular photo of a Northern Alliance tank driving through the streets of Kabul on its front page, noted in its caption that "A part of the population has welcomed the Northern Alliance soldiers as liberators" (November 14, 2001). Despite these reservations about the "liberation" in writing, photographs of "normalization" were published widely. These focused on the reopening of schools, the safety zones under U.S. control, the return of football, *buzkashi*, cock fights, dog fights, cinema, and other forms of entertainment, the release of prisoners from Taliban prisons, the shaving of beards, and the unveiling of women. But the fall of Kabul was not the end of the war. As the fighting increased, the newspapers turned more and more to reports of maiming, killing, and destruction. November headlines from the *Times* included "Refugees tell of frenzied killing in besieged city," "Beheaded hostages" (November 19, 2001), "Taliban chiefs vow a fight to the death" (November 22, 2001), "City of blood and betrayal" (November 23, 2001), "Dead litter city after frenzy of executions" (November 27, 2001). Pictures of wounded civilians, dead soldiers, and executions accompanied the written reports. The *Guardian* concluded on its front page "America Unloved" (November

24, 2001). A cartoon in its November 30, 2001 issue condemned the bombing of the Qala-e Jangi prison, where hundreds of people had lost their lives, and which *Le Figaro* described in terms of "apocalyptic scenes" (November 29, 2001). *De Morgen* published Robert Fisk's article from the *Independent* that probably summed up the feeling of many critics: "Now we are the war criminals" (November 30, 2001). The sympathy for the United States that was characteristic of the first days following 9/11 had completely dissolved by the end of November 2001.

The following month, the action moved to the Tora Bora cave complex and suspected hideout of bin Laden. The Bonn conference was on its way, and many believed that bin Laden would soon be captured. The press began to round up its narrative. The *Times* carried a close-up of an Afghan man on its front page of December 6, 2001, captioned "Face of hope: An Afghan waits for the new regime, and for food aid, in Kabul, yesterday." On December 22, 2001, the *Guardian* concluded "Liberated—Life begins to return to normal in Afghanistan" above a photograph of a public dogfight in Mazar-e Sharif. New photographs of normalization were published, and images symbolizing international friendship entered the press. The *Guardian* published a photograph of a New York police officer with orphans from Kabul and a picture of the Blairs on holiday in Egypt to indicate that stability had returned. The prime minister's job as war leader seemed finished at that time but a fortnight later the newspaper headlined "Day 100: another raid in the bombing war without end," and it added, "The Taliban may have vanished but the conflict is far from over for many in Afghanistan" (January 15, 2001). It now seemed that the story could not be ended after all. Nevertheless, media interest in Afghanistan began to decrease from 2002 onward. With the Taliban regime removed, for most of the media, the story of the Afghan intervention had been told. The tone toward the U.S. president and his administration became more favorable. The French newspaper, *Le Figaro*, concluded: "After arrogant beginnings on the international scene, the U.S. president has succeeded in rallying the world behind him" (January 19, 2001).

The number of photographs rapidly decreased after early 2002. A year later, when the media turned their attention to the Iraq War, Afghanistan was no longer seriously on the media radar. The deployment of Western troops, the international conferences on Afghanistan, and the military operations and foreign casualties still attracted attention. But the Afghan population, hardly ever at the center of the news, had now virtually disappeared from it. Whereas the intervention was still an international news story in October and November 2001, once European troops had landed on Afghan soil, western European newspapers turned their attention to their own national troops, and the debates surrounding their deployment, their numbers, and their equipment. Embedding further encouraged this domestic military perspective.

## THE FRAMING OF COMBATANTS AND TECHNOLOGIES

With no direct access to Western troops and the combat zone during the initial stages of the intervention in Afghanistan, telling the story of the soldiers in Afghanistan at first constituted a serious challenge. Photojournalists chose to make distant views of missile launches, airplanes, and air strikes instead. Combat photographs of the initial intervention were mostly of Northern Alliance troops on their way toward Kabul, Kandahar, and other cities. These early images rarely included civilian bystanders, and the enemy was implied only.[14] The first pictures of European troops sent to Afghanistan were noncombat photographs made at an airport or in a camp. Western troops figured mostly in photographs of activities such as war preparations and troop movements. When the practice of embedding was installed, photographs of their combat entered the press as well.

Initially, newspapers presented the intervention in Afghanistan as a tale of two wars: a high-tech battle of rational Western soldiers whose precision bombings caused collateral damage only, and a low-tech war of primitive warriors with a savage reputation. Among the first images of American high-tech power were those of aircraft carriers. Photographs of aircraft and aircraft carriers in the *Guardian*, *Le Monde*, *De Morgen* and the *Times* signaled and anticipated the impending war. The captions in the *Times* framed U.S. weaponry as superior and took over the language of the military. Emphasizing U.S. power, the *Times* described the aircraft carrier USS Theodore Roosevelt on its front page as "at the head of a powerful naval battle group" (September 20, 2001). The photo caption mixed the military with the personal, contrasting "emotion" with the "duty to carry out a mission."

Images celebrating military capabilities legitimated the war, reassuring the public that the war in Afghanistan was a low-risk enterprise for Western soldiers. Growing numbers of Western casualties later made this *managerial* frame increasingly unacceptable, but some newspapers already expressed doubts at the start of the operation. Photographers occasionally demonstrated the fragility of modern technology, for instance, with a picture of a helicopter crash (*Guardian*, November 7, 2001). In *Le Figaro*, a picture of the aircraft carrier USS Enterprise was accompanied by a caption that framed the U.S. army as superior, but it also questioned its effectiveness in the struggle against terrorism: "On board the aircraft carrier USS George Washington, Saturday, in the Atlantic Ocean. Trained for conventional wars, the most sophisticated and powerful army of the world is unsettled by the emergence of small invisible, diffuse and unpredictable terrorist groups" (September 17, 2001).

However, most photo-texts displayed an unambiguous fascination for war technology. Images of war technology focused attention on the *means* but not the *consequences* of war. During the first days and weeks of the military campaign in Afghanistan, when access to battlefields and troops remained limited, these images served as alternative visual evidence of the ongoing war. Newspapers published pictures made by military photographers, and caption writers frequently borrowed the military jargon. Words such as "target," "firepower," "military strategy," and "planning" were frequently used to accompany photographs of technological devices, eclipsing the human condition of war and representing it as a military-technological operation. On October 9, 2001, for instance, the *Times* published a photograph of a U.S. bomber plane taking off from its British base. The photo-text visualized the close relation between the United Kingdom and the United States and described the aircraft in military-technical terms, indicating that it carried "GBU28 "bunker-busting bombs." Bombing technologies were explained at length, often without any specific reference to the kind of human and ecological damage caused.

War journalists have been attracted by the technologies of warfare for a long time. Caroline Brothers (1997) described the fascination for the new war airplanes during the Spanish Civil War, and how the introduction of war aircraft radically altered our perception of the wartime landscape. According to Brothers, the most important effect of aerial photography was the introduction of an abstract, distant, and detached view. Aerial photography tends to dehumanize war, reduces human settlement and activity to landmarks, and removes indicators of human suffering. In aerial views, the people disappear and even the rubble might no longer be clearly observable. Texts added to aerial photographs contribute to this dehumanized perspective when targets of air strikes and operations are labeled as "areas," "sites," or "targets." Buildings are indicated rather than people: "TV tower," "Defense Ministry," "Airport." The local on-the-ground realities are eclipsed in favor of abstract dots that mark buildings, explosions, cities, roads, and so on; the viewer is invited to think as a soldier and interpret events in a military-tactical way. Additionally, aerial pictures produced by satellites without direct human intervention suggest an omniscient perspective that is more objective than human observation.

In the news, war technology does not merely appear as the means to conduct war. War technologies regularly become grammatical actors in the ongoing battle, both textually and visually, replacing or obscuring the presence of human beings. In a caption accompanying a photograph of an aircraft published by the *Guardian*, it is the *gunship* that targets troops, not those inside the gunship: "The AC-130 Spectre, a converted transport plane that can fill a target area with machine gun bullets in seconds as it slowly circles ... some analysts have pointed out its victims will include pressganged Afghan conscripts." This form of technologization and objectification deflects human responsibility, especially when the target or goal of the action is kept out of sight. The *Guardian* did include a reference to the human damage made by the aircraft—it kept the human cost on the agenda, especially in the poignant cartoons of Steve Bell—but other newspapers regularly left out the victims in photo-texts of war technologies.

The dehumanization of war through an emphasis on weaponry appears to be a widespread phenomenon that occurs across the boundaries of public affairs, PSYOP, and the media. For example, a picture of an AC-130 Spectre aircraft, almost identical to the one used in the *Guardian*, was used on a PSYOP leaflet to intimidate the Taliban and Al-Qaeda combatants (Fig. 3.3) but with the leaflet's text directly addressing the enemy: "Taliban and Al-Qaeda Fighters—We know where you are hiding." Similar images can be found in DoD public affairs photo collections.

**Fig. 3.3. "Taliban and Al-Qaeda Fighters—We know where you are hiding" (PSYOP leaflet AFD40d).**

An almost identical image was published in the *Guardian* on October 17, 2001 (page 3, "Devastating gunships target troops").

Photos of war preparations began to appear shortly after 9/11. *Le Figaro* (September 17, 2001) published a picture of the preparations on the USS George Washington aircraft carrier; *De Standaard* (September 21, 2001) published a photograph of soldiers exercising at Fort Campbell; and the *Times* (September 21, 2001) carried a photograph of troops arriving at a military airport in Oman. The pictures suggested the imminence of war and the foreclosure of peace negotiations.

Western soldiers appeared as sensitive, united, orderly, and disciplined. An article in *Le Monde* headlined "The Marines against the 'chaos'" (September 29, 2001) opposed the dedication to order of the

Marines to the supposed anarchy of Afghanistan. In *Die Welt* and the *Times*, photographs of soldiers offered glimpses into their daily life. The *Times* published a photograph of soldiers eating lunch with the caption: "Crewmen eat lunch on the flight deck of the *USS Enterprise.* Pilots from the carrier have destroyed Taliban fighter jets, its officers claimed" (October 10, 2001). The caption normalized the violent ("destroyed Taliban fighter jets") by associating it with the mundane ("eat lunch"). *Die Welt* (November 2, 2001) published a similar scene from daily life, two Marines celebrating Halloween. The photographs framed Western soldiers as "like us," making the appearance of combatants and war more acceptable. The *Times* portrayed them as poetry lovers in an article entitled "Dead poet's society eases the tension of war" (November 1, 2001). The caption below the photograph of a soldier looking out at sea was: "Easing the tension: a Marine greets the dawn on the flight deck of USS Peleliu in the Arabian Sea. Many of the crew are also studying First World War poets like Wilfred Owen." The focus on the soldiers' stress made them more humane than their Afghan counterparts, who were usually stereotypically framed as fearless and "willing to die." The association with British poet Owen, who was killed in action in 1918, established a link between World War I and the intervention in Afghanistan. The reference suggested a view of history as a sequence of wars. It implied that Western soldiers in Afghanistan were facing similar perils as those confronted with trench and gas warfare during World War I, even though, during the initial high-altitude bombing campaign, Western soldiers were hardly at risk and their situation was not comparable to that of troops in World War I. A comparison with World War I, in which a massive number of soldiers died, has a strong emotional appeal, especially to the British who continue to commemorate their fallen soldiers in Flanders Fields. The reference to Owen, who rejected Horace's *Dulce et decorum est pro patria mori* (*It is sweet and right to die for one's homeland*) as "the old lie" in his famous poem "Dulce et decorum," also suggested that British soldiers were as anti-war as the poet. Thus, at the onset of the intervention, the reference created a positive image of the combatants, not willing but *obliged* to fight. In the *Times*, images of the arrival and departure of soldiers served the same purpose. The soldiers were going to war out of necessity and moral duty. They were motivated by external factors, not because they had some violent internal dispositions as Afghan combatants did. Although Afghan and Western soldiers were caught up in the same acts of violence, the right intention of the latter made their act morally acceptable, it was suggested, and this reasoning helped to establish the intervention in Afghanistan as a just war.

Photographs of the departure and arrival of soldiers emphasized the efforts that had to be made by families who were split apart by the war (e.g., in the *Times*, November 12 and 15, 2001). Pictures of departure visualized the temporary breakdown of the family. Those who departed were described as serving the higher-order unity of the nation. Arrival scenes depicted the reunion of soldiers with their spouses and children, and affirmed the unity of the group.

Western soldiers also were framed attributively as techno-combatants. These images celebrated rather than questioned technological advancements, stressing the military capabilities of technical devices. But in the build-up to the intervention and its early stages, newspapers seemed to take more visual interest in war technologies than in Western soldiers, perhaps also because of the lack of access to troops. One striking feature of the early combat and combat-related images of Afghanistan is the absence of strong emotions for which many iconic war photographs are known. Vietnam photographers captured the emotional distress on the face of soldiers who were going through the war. Photographers like Larry Burrows, whose picture of a distressed helicopter crewman facing the death of a young lieutenant was published on the cover of *Life* (April 16, 1965), acted as participant-observers. Their practice was reflected in how they portrayed soldiers: as human beings experiencing the horror of war. Initially, when photographers had no access to the combat zone, Western soldiers in Afghanistan were mostly depicted as technical operators of war machinery, as in many DoD images. Northern Alliance combatants often were depicted from a distance and from the back. Many photographs depicted soldiers "on the move" or preparing weaponry, thus creating an impression of the war as a logistical operation. Few photographers covered the war by depicting the dramatic experiences of individual soldiers as Burrows did in Vietnam. It was only after journalists gained access to troops that more emotional pictures began to appear.

Newspapers published some photographs of soldiers in interaction with members of the local population, but not very frequently, and those that did, mostly supported a propaganda view of the Western soldier as a friend of the local population and a peacekeeper. For instance, one photograph in the *Times* depicted a British soldier shaking hands with an Afghan man, symbolic of a hoped for relationship between the two countries. As with similar images used in PSYOP, the photo-text framed the arrival of international troops as something positively appreciated by the local population, with the caption identifying both participants in the scene as representatives of their nations: "A British soldier in the streets of Kabul yesterday, shaking hands with an Afghan man. Pedestrians waved at passing peacekeeping vehicles and one elderly man saluted them" (December 26, 2001).

Northern Alliance, Taliban, and Al-Qaeda troops were depicted as almost "naturally" violent. The contemporary representation of the naturally violent Afghan is part of a longer tradition of representing

Afghanistan as an inhospitable and unconquerable terrain, and of locating the roots of Afghanistan's problem inside the country, going back, in the case of the British, to their experiences during the Anglo-Afghan Wars of 1839-1842, 1878-1880, and 1919 (Fowler, 2007). The presumed anarchy and violence of Afghans is represented as an aspect of their nature, whereas the violence of the West is related to external situational factors (i.e., necessary due to the threat of terrorism). This representational pattern can be seen as an illustration of actor-observer bias. According to the theory of actor-observer bias, observers are likely to attribute the behavior of others to internal dispositions (Lee, Hallahan, & Herzog, 1996).

The assumed violent and anarchic nature of the Afghans was interpreted as a sign of their irrationality. For example, journalists found it fascinating and surprising that Northern Alliance and Taliban members could change camps quickly and easily.[15] Rather than interpreting this as a rational survival strategy, the press often saw such events as signs of incomprehensible Afghan irrationality.

The assumed violent nature of Afghan combatants was linked to Islam through visual and textual co-location. The *Times* published a photograph of a soldier during his prayers with the caption, "An Afghan opposition fighter prays near the village of Marathaja, about eight miles from Kabul, yesterday" (October 13, 2001), and of a praying soldier to accompany an article called "Battlefield of guns and prayers" (October 27, 2001). The picture was captioned: "A young Afghan at Friday prayers in the mosque in Khoja Bawaddin, northern Afghanistan, with two AK47 rifles resting on his prayer mat." *Die Welt* captioned a similar photograph of praying Taliban as "Allah gives the marching orders: Taliban-fighter at the Friday prayer in Kabul's largest mosque. The weapons of the religious warriors are within reach" (November 3, 2001). These photo-texts created a strong association between violence and Islam, reinforcing the idea of Islam as a threat to peace and promoting religion as a causal frame. Several other photo-texts in the western European press suggested that Muslims were "raised" for war. The *Times* illustrated an article by reporter Catherine Philp, called "Mohammed, 10, is ready for war" (October 3, 2001), with two portraits. One was entitled "Mamat Sallari, 15, will fight with the Taliban," the other "Mohammed Raoul is committed to the jihad." Another article by Philp developed the same theme. It was headlined "Brutal message lives in minds of the children" (December 17, 2001). The cliché of youngsters brainwashed to participate in jihad returned in images of boys studying the Quran in madrasas, schools that became synonymous with fundamentalism in 2001.

Children also were represented as lonely victims, removed from their family and their everyday environment. Images of isolated children under the protective care of Western soldiers or aid workers legitimated the intervention. Photo-texts of physically wounded, socially abandoned and mentally corrupted children contrasted strongly with the Western iconic image of the child as innocent and playful. They justified the war by attributing causal responsibility of the assault on children to the Taliban. For example, a photograph of a father and his child in the *Times* was captioned: "Abdul Fatiloh, 4, whose leg was shattered by the Taliban because his father, Nurala, had escaped from them" (November 12, 2001). Another image in the same newspaper was captioned "Goldinah, 20, who was raped and beaten when 30 Taliban broke into her home" (November 12, 2001). Captions attributed problems of Afghanistan to the Taliban and their supporters without a consideration of the geopolitical factors that produced and sustained their practices. Even when articles offered a more subtle in-depth analysis, few captions provided contextual information about the Deobandi ideology of the Taliban movement or the role of Pasthunwali and Wahhabism.[16] Captions functioned as arguments in favor of the need to intervene, depicting the Taliban as the root cause of Afghanistan's problems. With relatively few images providing visual evidence of their abuse, most newspapers evoked Taliban extremism indirectly. One photograph in the *Times* used "the museum's destroyed collection" (November 23, 2001) in Kabul as a metonymy for the complete destruction of Afghan culture by the Taliban.

Like Northern Alliance soldiers, Taliban combatants were endowed with a violent disposition. A photograph in *Le Figaro* of Taliban prisoners, for example, was captioned: "Taliban prisoners in Panshjir. Most of the fundamentalists are ready to die for their cause" (September 19, 2001). An article in the *Times* framed the Taliban as revengeful: "Refugees tell of Taliban revenge for U.S. bombing" (November 8, 2001). The cause of the refugee problem was defined in the newspaper, not as the bombing itself, but as the Taliban's response to it: "Ayat Tali, one of the refugees who *fled from the Taliban terror* in Dahse-i-Archi, with his three year old son, Juhamon, at the Khojama camp" (italics added).

From November 2001 onward, the Taliban were not only depicted as perpetrators but as prisoners and defectors as well. But it was the uprising at the Qala-e Jangi fortress near Mazar-e Sharif that led to a more dramatic turn in framing. The uprising of mostly foreign combatants who had joined the Taliban took place between November 25 and December 1, 2001. The combatants had surrendered to Northern Alliance General Abdul Rashid Dostum after the bombardment of Kunduz. Using weapons that had been smuggled into the prison, they created a riot after two prisoners had blown themselves up. To end the uprising, U.S. Special Operation Forces and British Special Air Service commandos backed up Northern Alliance forces with heavy air strikes. The battle eventually led to the death of hundreds of Taliban, approximately 40 Northern Alliance soldiers, and the first American to die in combat in Afghanistan, CIA agent Johny Michael Spann. Journalists present at the site became part of the conflict. A *Time* translator, Nagidullah Quraishi, was ordered to translate communications between the Western soldiers and their Northern Alliance allies, and the satellite phone of the German television network ARD-TV was used to call the U.S. embassy in Tashkent for help.

The *Guardian* called Qala-e Jangi "the defining moment" and compared it to the 1968 My Lai massacre in Vietnam and "the remains of charred Iraqi soldiers on the Mutla ridge outside Kuwait" (December 1, 2001). The newspaper saw the treatment of the Taliban in the prison—some died with their hands tied behind their back—and the firepower used to control the uprising as a breach of the Geneva conventions but U.S. and U.K. officials dismissed these accusations. The Taliban were temporarily framed as victims, and the uprising had an impact on the framing of the Northern Alliance soldiers too.

Photographs of Northern Alliance soldiers had been published well before the intervention of October 7, 2001. They were depicted in surveillance activities and war preparations. In accordance with the colonial trope of the "wild" Afghan, and almost from the start of the intervention, captions regularly referred to their "savage" nature, particularly in the *Times*. An article below a photograph of Northern Alliance troops called them "thirsty for Kabul and blood of Talebs" (*Times*, October 5, 2001). One headline, consistent with the idea that unruly Afghans pose a threat, called the Northern Alliance a "dangerous ally" with a "history of slaughter" (*Times*, November 12, 2001). Captions framed the Northern Alliance as "wild" and "primitive" warriors, who also enlisted children. But before the fall of Kabul and the uprising at Qala-e Jangi, there was a discrepancy between these textual frames and the visual content of photographs of the Northern Alliance. The photographs did not associate Northern Alliance soldiers with any overtly brutal war activity. Visually, the Northern Alliance troops appeared as low-tech soldiers, moving around, preparing weaponry, waiting, or firing at a distant enemy. Northern Alliance soldiers were frequently depicted as observers of the U.S. bombing campaign rather than as participants.[17] This changed in late November 2001, after Qala-e Jangi. *De Morgen* published a photograph of a man threatened by Northern Alliance soldiers in the streets of Kunduz (November 27, 2001), and the *Times* carried a photograph with Northern Alliance soldiers that according to the caption "paraded" Al-Qaeda combatants "for ritual humiliation" (December 17, 2001). Newspapers increasingly began to question the role of the Northern Alliance. Previously still appreciated for their role in the liberation struggle, they were now seen as violators, obstructing the development of a free Afghanistan. Throughout the entire period, photo-texts attributed the roots of the violence in Afghanistan to the presumed brutal character of the Afghan in ways that appear largely consistent with 19th-century colonial ideas. But the supposedly violent nature of the Afghan was not simply despised. The Afghan soldier remained a source of admiration and fascination too, even after Qala-e Jangi. Apart from repulsion about Afghan "savagery," epitomized in the Afghan warlord and brutal Taliban, journalists were attracted by the fearlessness and bravery of the Afghan combatant. Before the intervention and during its first weeks, journalists liked to remind their readers that Afghanistan had never been conquered before, neither by the British Empire in the 19th century nor by the Soviet Union in the 20th century. *De Standaard* was one of many to note that "Many foreign conquerors were defeated [in Afghanistan]" (September 17, 2001), and the *Times* published an old woodcutting depicting a scene from the Great Game to illustrate an article that was entitled "Deadly warrior tribes kept invaders at bay" (September 15, 2001). The Great Game brought to mind Rudyard Kipling's popular writings, especially *Kim* (1966) and *The Man Who Would Be King* (1987). References to Kipling were widespread in the U.K. press and created an image of Afghanistan as a country of danger and doom that is difficult to separate from British experiences during the Anglo-Afghan Wars, and to the failure to incorporate Afghanistan into the British Empire.[18] The Kiplingesque perspective traveled across the Atlantic. As Robert Young Pelton (2002) reported for CNN: "When Dostum would roll into a city to liberate it, it was almost Kipling-esque, because you'd see Dostum with his turban, the warlord, and then you'd see these two guys wearing turbans and Oakley sunglasses sitting next to him, the CIA. So, it's almost like 'The Man Who Would be King' you know" (para 20).

Attracted by the supposedly wild and brave nature of the Afghan, Western soldiers occasionally swapped identities and became "one of them." A photograph published in the *Times* of December 17, 2001 (Fig. 3.4) depicted a U.S. soldier during a game of *buzkashi* (literally "goat pulling"). The elaborate photo caption is worth looking into:

> **American grabs a piece of the Afghan action**
> An American serviceman is caught up in Afghanistan's newfound freedom, and the brutal national sport of buzkashi in Mazar-i-Sharif, yesterday. ... Mazar residents flocked to watch about 50 buzkashi horsemen fighting for control of a headless goat. Players drag the corpse around the arena and try to deposit it in a goal. Yesterday's victors won prizes of wristwatches and cash of up to £15—a significant sum in Afghanistan. When Hamid Karzai was chosen to head the new post-Taliban administration in the country, he said he dreamt of the day when Afghans could hold a free loya jirga, or traditional assembly, and then celebrate with a wild game of buzkashi.

**Fig. 3.4. American buzkashi player. © Maxim Marmur/ A.P./Reporters**

Published in the *Times* on December 19, 2001 (page 12, "British troops join Afghan patrols in Kabul").

*Buzkashi* is described in the caption as "brutal" and "wild," and the residents do not simply come together to watch the game, but "flock" to it. The U.S. soldier does not participate in the game but "grabs" a piece of the action and is "caught up" in it. The facial expression of the U.S. soldier looks brutal and fierce too, indicating that he has become a "wild" rider himself. The image seems to tap into the familiar American archetype of the cowboy, contributing to what Fowler (2007) called "the Wild Westification" of Afghanistan. The representation of *buzkashi* as a wild, brutal sport belongs to the popular Western imagination, and is found in films from the Cold War era: *The Horsemen* (Frankenheimer, 1970), *The Man Who Would be King* (Huston, 1975), *High Road to China* (Hutton, 1983), and *Rambo III* (MacDonald, 1988). The reference to wristwatches and cash in the caption of the *Times* suggests that the game is played for material gain. But according to Azoy (2003), *buzkashi* is a complex cultural ritual that has allowed leaders to "play out" their disputes and determine power without resorting to actual warfare. In more recent times, its nationalized version, *qarajai*, has been used in an attempt to construct a cohesive national identity, including by President Hamid Karzai. In photo-texts—and Azoy's work as well—the game tends to be reduced to an all-encompassing symbol of Afghanistan although it actually occurs throughout central Asia, in various forms and with multiple meanings.[19] The press used *buzkashi* as a symbol of what it means to be Afghan and to stress how much Afghans differ from "us."

*Buzkashi* pictures were sometimes mere forms of exoticism, but the photograph in the *Times* can be read in multiple ways. First, the *buzkashi* player in this image is a symbol of inversion: The player is not Afghan but American (Babcock, 1973). The supposedly rational techno-warrior descends into the realm of the supposedly primitive, bodily, violent, and chaotic world of the Afghan warrior. Retaining some features of his original identity (sunglasses and uniform), the U.S. soldier appears as a hybrid. The resulting heterogeneous, unstable space is an opportunity for the viewer—seized or not—for questioning the mainstream bipolar portrayal of U.S. and Afghan soldiers.

In Barbara Babcock's work, symbols of inversion function as paradoxes that are capable of questioning established orders but, in accordance with Turner's (1967) view of the liminal stage in traditional rites of passage, she also saw them as necessary ways to reproduce these orders. In her structuralist analysis, order needs disorder to sustain itself, as anti-structures in Turner's work ultimately function to support societal structures. Following this logic, representations in which U.S. soldiers enter the disorderly Afghan world ultimately affirm the order to which these soldiers belong. For some, the image of the American *buzkashi* player might indeed express and induce a fear for the return of what is repressed, and signal a potential breakdown of the community. The viewer is confronted with a sort of "fall" of the subject into abjection. The open mouth of the American *buzkashi* player can be seen as a cave or hollow that symbolizes the soldier's descent into the world of the body and the flesh that is so important in Kristeva's work on the abject, and that is further evoked in the picture by the goat, the horse, and the physical posture of the horse rider. Otherness is no longer simply an outside. It is inside the American *buzkashi* player; it is embodied.

The American *buzkashi* player appears to be negotiating his identity with the other, with the supposedly primitive nature of the Wild Afghan. The image seems to suggest a suppressed savage irrational side and may function as a warning sign. But the American *buzkashi* player can also be read as an instance of mimicry. Mimicry is usually seen as an instance of ambivalence that occurs when the colonized subject is encouraged to imitate the colonizer, for example, by accepting the colonizer's dress codes. Bhabha (1994) argued that mimicry is a source of anxiety for the colonizer when it becomes a parody and an appropriation. It then threatens to undermine colonial authority. But many other interpretations of mimicry exist. The *buzkashi* player is not unlike Kipling's Kim, who is presented as someone who dresses like, and could pass for, an Indian. For McClintock (1995), Kim uses racial mimicry "not to subvert colonial authority but to enhance it" and he is described as "an allegory for a reformed and more discreet style of imperial control" (p. 70). The *buzkashi* player in Fig. 3.4. can be understood in a similar way: The U.S. soldier has joined the fight of Afghans to better control it. He is presented as a privileged player, capable of transcending the boundaries of his own culture. The goat that the game is all about is on *his* horse, which suggests that the U.S. soldier is even better at *their* game.

Ambivalences in the portrayal of combatants appeared elsewhere too, but with different connotations. Westerners who had joined the Taliban or Al-Qaeda intrigued the media. Among the survivors of Qala-e Jangi were two U.S. citizens fighting with the Taliban: Yaser Esam Hamdi and John Walker Lindh. "American Taliban" John Walker Lindh had come to Afghanistan via Pakistani groups that were affiliated with the Taliban and Al-Qaeda, and was detained at Camp Rhino, southwest of Kandahar. His uncertain status-—he refused to reveal his identity—was an important news frame. His story apparently struck a chord, with images of him appearing outside the press as well. Artist Jon Haddock made a small statue in polymer clay that made Lindh look like a crucified Christ, a prophet who informed others about future times to come. In accordance with the status of a prophet, press photographs and captions of images of Lindh emphasized his marginal and controversial position. A key element of this frame was a focus on his bodily appearance: He was depicted as a dirty, wild man, with uncombed hair and beard. Whereas the American *buzkashi* player in Fig. 3.4 can be valued positively as strong and fearless, the Western Taliban were portrayed as negative impurities of the system. Their dirty looks suggested the need to be "pure."

In her classic work, *Purity and Danger*, Douglas (2002) argued that dirt is a symbol for the placeless, marginal and incompatible, and therefore also for the morally wrong and dangerous. Seen from this (structuralist) perspective, images of the "Western Taliban" operated ritualistically to warn viewers of what would happen if the established East–West order was offended. To the extent that such a perversion is perceived as a threat to the established order, these images call on the viewer to reinforce it and to purify it from defilement. Presented as physically dirty and morally inferior, Lindh appeared as a threat to the community and his actual expulsion from the community was welcomed as a purifying ritual and became a major news event.

Images of dead soldiers represent the death of the normal order but the military has been quite successful in keeping representations of corpses out of the news (see also Louw, 2003). In the early years of the conflict, the death of Western soldiers was represented indirectly only, as in a picture of a grandparent and grandchild visiting a graveyard (*Die Welt*, October 25, 2001) or a photograph of U.S. soldiers loading a coffin unto an aircraft (*De Morgen*, January 17, 2001). In the latter case, the photographer used blurred motion, which gave the scene an abstract quality.

Dead foreign combatants became major news when the uprising at the Qala-e Jangi prison occurred. Graphic images of the violent suppression of the uprising appeared in all newspapers, particularly *De Morgen* and the *Times*. *Le Monde*, *De Standaard*, *Die Zeit*, the *Guardian*, and *Die Welt* were more reluctant to show scenes of blood and gore. *De Morgen* carried a large color photograph of two dead Taliban soldiers in its news supplement “Reporter.” The caption described the photo as “The bodies of slaughtered Taliban fighters, sometimes shot in the back of the neck, in a pit near the Qala-e Jangi fort. The hands of the combatant at right are tied” (December 1, 2001). Words such as “body,” “slaughter,” “back,” “neck,” “pit,” and “hands” all stress the physical and bodily. The social and psychological aspects of death were excluded.

*De Morgen* published other photographs of the Qala-e Jangi uprising. The photo on the front page of its November 29, 2001 issue depicted a Northern Alliance soldier taking away the shoes of slain Taliban soldiers. The caption universalized the event, and raised doubts about the conduct of the forces, although without any reference to specific perpetrators: “The stripping of corpses is of all wars. A soldier of the Northern Alliance collects shoes of the better class of dead Taliban fighters near the fort of Mazar-e Sharif, in North-Afghanistan. In and around the fort an uprising of Taliban and Al-Qaeda fighters raged for days and cost hundreds of people their lives. Yesterday, Amnesty International demanded an independent investigation into the facts that occurred in the Qala-e Jangi prison.” On November 17, 2001, *De Morgen* devoted an entire spread to photographs by Luc Delahaye, including two pictures of slain Taliban. For a short time, the Taliban became victims whose suffering was treated in detail. Their death symbolized the end of the Taliban regime.

## VISUALIZING LEADERSHIP IN WARTIME

Photographs of heads of government occupy considerable space in any newspaper at any time. The pictures personalize the news, and create an image of a world run by a small number of individuals, who seem to bear responsibility for all that happens. Especially in times of war, heads of governments are not only executives of their nation and representatives of the national public. They *lead* and *are* the nation. Political leaders may appear individually, in the company of other leaders, or, more rarely, among their own citizens or soldiers. Depending on who is in the photograph, pictures of the meetings and greetings of leaders frame war as a national or a global event. Who is depicted with whom, and at what distance, is symbolic for the kinds of international relationships that exist. In 2001, for example, close-ups of Gen. Pervez Musharraf with Western leaders visualized new intimate international alliances with Pakistan.[20] A meeting of President Bush, Chinese President Jiang Zemin, and Russian President Vladimir Putin to gather support for Bush's operation in Afghanistan also featured prominently in the press, but they were visually represented from a longer distance. Such variations in the representation of political leadership may simply result from situational constraints but they also seem to be determined by the political orientation of the newspaper. They indicate which international relations are deemed important, the extent to which an event is considered of international importance, and the political orientation of the newspaper. In the immediate post-9/11 period, Belgian newspapers focused on the role of the European Union (EU), and to a lesser extent on that of their own leaders. German newspapers focused on Minister of Foreign Affairs Joschka Fischer of Germany's green party and prime minister and social democrat Gerhard Schröder. The domestic orientation in Germany partly resulted from a dispute between Fischer, who supported the military action, and his party leader Claudia Roth, who called for a halt in the bombing campaign to allow for humanitarian intervention.[21] In *Le Monde*, pictures of leaders of the United States, the UN, and the EU invited readers to take a more global perspective. In the run-up to the intervention in Afghanistan, and in accordance with its international orientation, *Le Monde* published photographs of French President Jacques Chirac with UN leader Kofi Anan, and with the then president of the European parliament, Nicole Fontaine. *Le Monde* also published a close-up of Blair with a headline that framed him as "a war leader" (October 25, 2001), and a photo of Blair amidst his troops on the inside. Such images in which leaders appear among their troops typically support the idea of war as a common cause, and meetings with soldiers are often staged as photo opportunities to promote such a frame. The images in *Le Monde* seemed to suggest that the British were ready for war.

*Le Figaro* put more emphasis on U.S. leadership and U.S.–French relations, although the newspaper also published a photo of NATO leaders. The newspaper took a more pro-American stance than *Le Monde*. It gave prominent visual space to Colin Powell and National Security Advisor Condoleeza Rice, although after the start of the intervention *Le Figaro* also published photographs of Schröder. *Le Figaro* framed Chirac as a friend of the United States, and of Bush in particular. A photograph of Chirac and Bush in Washington was published on the front page of *Le Figaro* on November 20, 2001. A photograph related to the same meeting depicted Jacques Chirac and French Minister of Foreign Affairs Hubert Védrine leaving the White House, and the caption described the meeting as "friendly." Two weeks earlier, *Le Figaro* had published another photograph of the French and U.S. presidents, a close-up symbolizing their friendly relations. Inside the newspaper, however, another picture of the two presidents was captioned in a way that suggested a possible divergence between the French and U.S. views: "Jacques Chirac, yesterday in Washington, next to George Bush: 'We have reaffirmed our solidarity with the Americans, all the while being conscious that the military action, indispensable as it is, *is not the only way to combat international terrorism*'" (November 7, 2001, italics added).

This position was shared elsewhere. "It is not the only way to combat international terrorism" was a line picked up by the *Guardian*, the British newspaper that was far more critical of the war than the *Times*. The *Guardian* focused on Blair's role in Europe, publishing photographs of Blair with Chirac, with Chirac and French Prime Minister Lionel Jospin, and with Germany's Prime Minister Schröder. The *Times* gave Blair a more global profile and presented him in more diverse roles: as a military leader, a family man, a diplomat and negotiator, a peacemaker and a world leader, and a friend of the United States. The photo-texts emphasized his capacity for creating unity within and across the nation. The visual framing of Blair as an important world leader and diplomat mainly occurred through photographs of his meetings with a host of elite leaders during the preinvasion period. On September 22, 2001, the *Times* published three different photographs of Blair: one with Chirac, one with Bush, and one with Italian Prime Minister Sergio Berlusconi. The newspaper also published a photograph of Blair next to U.S. first lady Laura Bush during the U.S. president's September 20, 2001 speech, receiving applause from her and members of the U.S. Congress. The headline of the story was "On board the Blair flight of diplomacy." The page title, "America at war: The Blairs," counter-framed the headline's emphasis on diplomacy. The page's streamers did so too: "As he swooped from one destination to the next, the language of war grew ever graver," and "Rush-hour traffic is slowed by commuters: don't they know there's a maybe-war on." Although Blair was occasionally represented as a peacemaker, diplomat, and in other roles, the predominant frame was that of a war leader, consistent with the initial overall war-oriented reporting of the *Times*. On October 6, 2001, the *Times* gave Blair epic dimensions in a section called "War on Terror: Blair's Odyssey." The newspaper carried one photograph of Blair on board his airplane, and another one of Blair with President Musharraf from Pakistan. The latter symbolized the rapprochement between the West and Pakistan, and the key role of Musharraf in the Afghanistan conflict at that time. A similarly structured photograph appeared on October 17, 2001, of Musharraf and Colin Powell.

Blair was also depicted in the company of his spouse, Cherie Booth. "The Blairs," as they were called in the *Times*, to stress their unity, became the ideal-typical wartime family in that newspaper, a model for the nation: caring for each other but ready to act for the greater cause too. The royal family, a model of British unity in earlier wars, was visually absent in the press, with the exception of some pictures of the prince of Wales.

As a model for families and the nation, the Blairs called on the public to make the necessary sacrifices, in a way reminiscent of total war: "Blair says good life must be put on hold" (*Times*, November 15, 2001). The portrayal of both reinforced traditional roles. Blair was framed as ready to exercise hard power to defend the nation while his wife was framed in a soft-power role supporting the "liberation" of Afghan women. She assumed the same traditional and supportive role as Laura Bush did in the United States. When the war had started, the emphasis shifted entirely to Blair as the competent military leader. On October 11, 2001, the *Times* headlined, "Blair defines his war aims" and published an image of the prime minister amidst his troops. Inside the newspaper, another photograph served to break down the hierarchy between Blair and the common soldier. The picture of Blair serving himself an ordinary soldier's lunch suggested an atmosphere of communality. The text accompanying the photograph of Blair and his troops revealed perceived differences between the U.K. and U.S. approaches. The photo caption quoted Blair who "said his immediate war aims did not go beyond Afghanistan," while a subhead on the same page said "Bush looks to a wider conflict." Other newspapers too indicated that European leaders differed from their official U.S. counterparts. They reported disagreement over the *extent* of the conflict, as in the *Times*, and disagreement over the *means* to resolve the conflict.

By the end of November 2001 the *Times* considered the war largely over, and felt it was time for an assessment of Blair's performance as a war manager. The newspaper concluded: "Blair's wartime rating is higher than Thatcher's" (November 30, 2001). The composite picture that accompanied the article combined portraits of Blair and Thatcher with an army vessel in between them. It summed up the dominant vision of war as a technological operation managed by strong leaders. The appearance of Winston Churchill in the *Times* similarly served to frame political leaders as "war managers." Churchill was framed as the exemplary and inspirational leader, who could inform contemporary leaders on "How to manage a war" (September 25, 2001), as one headline on the newspaper's front page said.

The availability of appropriate equipment was presented as another precondition for managing war properly. Both frames were combined in images of political leaders with the tools of war. The *Times* of October 9, 2001, for example, published a picture of Geoff Hoon, the U.K. defense secretary, with a warplane in the background. The picture represented the war in an elitist way by focusing on Hoon, and in a technological way by showing a military aircraft in the background. The composition of the picture suggested a future of war as a technological operation managed by elite leaders: Both the secretary and the plane were "looking" in the same direction, to the right, which in Western visual culture, can be associated with the new and the future (Kress & van Leeuwen, 2003).

For the press, the two protagonists and adversaries in the war story were Bush and bin Laden. Following the 9/11 attacks in New York and Washington, all newspapers published photographs of the U.S. president and bin Laden. They represented metonymically the U.S. nation and (Islamist) terror, respectively. During the initial stages of the intervention, both were framed as opposites, and the opposition continued to exist even when the framing of both shifted. Between September 11 and September 20, 2001, Bush was framed as a vulnerable and emotionally shaken leader, whereas bin Laden appeared powerful and confident in most of the pictures. Bin Laden was later framed as the object of a manhunt, and gradually lost his image of a powerful terrorist. By the end of 2001, the frames were entirely reversed: Bush was represented as a powerful figure, bin Laden as vulnerable and weak. This portrayal of the protagonists was part of a larger narrativization of the intervention, meant to provide it with a clear beginning ("conflict"), middle ("fighting"), and end ("victory").

The press considered the president's speech before a joint session of Congress on September 20, 2001, as a veritable turning point. The speech dramatically changed the framing of the U.S. president. Prior to September 20, 2001, Bush's vulnerability was emphasized in such pictures as a close-up of his tearful eyes, published in the *Times*, *Le Figaro*, and *De Morgen*, and in pictures of his early September 11, 2001 speech, where he was shown biting his lips and with his eyes cast downward, published in the *Times*, *Le Figaro*, and *Le Monde*. The *Times* combined the framing of Bush as vulnerable with a call to act dutifully and decisively. The close-up of his tearful eyes, which the newspaper put on its front page of September 14, 2001, was captioned as follows: "Tears well in the eyes of President Bush as he speaks of the families and victims during an Oval office press conference. He said: 'I'm a loving guy, and I'm also someone, however, who has a job to do and I intend to do it.'" What the "job to do" entailed was indicated by the headline above the photograph: "America prepares for war." *Le Figaro* carried the same photograph on page 3, but framed it with an entirely different text that urged for calm rather than war: "George Bush incited his compatriots to treat Arab Americans and Muslims in general with respect."

After September 20, 2001, all newspapers published images of the president as a strong, confident leader. Several newspapers published the same photograph of Bush addressing Congress, a picture by Win McNamee. The low angle used by the photographer emphasized the president's power. His superordinate position was strengthened by his foreground position, with the Speaker and President Pro Tempore of Congress applauding him in the background. The inclusion of the words "In God We Trust" at the top gave the pending war a religious connotation: Bush was acting on God's behalf.

*Die Welt*, which carried the photograph on its front page, used a caption that repeated Bush's words: "George W. Bush during his historical speech in Congress—The U.S. president offered governments throughout the world a choice: 'Either you are with us or with the terrorists'." The headline, "Bush: it was bin Laden," further strengthened the dichotomy. *De Standaard* carried the same photograph but framed the pending war more narrowly as a conflict between the United States, the Taliban, and Al-Qaeda: "President Bush demanded Thursday evening from the Taliban the 'immediate hand-over' of Osama bin Laden and other leaders of Al-Qaeda. Otherwise the Afghan regime risks the revenge of the U.S." Whereas Bush was merely "looking forward to solidarity" (September 15-16, 2001) in an earlier caption in the newspaper, he was now more overtly presented as a demanding and threatening leader. The *Guardian* published the photograph too, but it cropped the lettering "In God We Trust," avoiding the religious connotation. It used a quote that characterized the president's speech, with a strong emphasis on the national community through the use of an all-inclusive "we":

> Our nation, this generation, will lift the dark threat of violence from our people and our future. We will rally the world to this cause by our efforts, by our courage. We will not falter, and we will not fail. (Bush cited in *De Standaard*, September 15-16, 2001)

Also focusing on the same inclusive "we, the nation," the *Times* used "We will not falter, and we will not fail" as a headline above another photograph of Bush (September 22, 2001). The earlier reservations about Bush made by the *Times* had now clearly disappeared. The newspaper published the signatures of Bush and bin Laden on its front page of September 25, 2001, below the headline "With the stroke of a pen, Bin Laden and George Bush declare war." The caption accompanying Bush's signature affirmed the new frame: "President Bush's signature shows determination." The caption accompanying bin Laden's signature turned the latter into a weak character: "The signature of Osama bin Laden on yesterday's statement: it shows someone who is chronically unhappy and in great need of self-protection, according to a handwriting expert." The images portrayed the conflict between the United States and Afghanistan in an entirely personalized way, as a clash between personalities: one stable, the other neurotic.

By mid-November, when Mazar-e Sharif had fallen, photographs depicted a triumphant U.S. president. *Le Figaro* published a picture of a smiling president giving the "thumbs up" sign, and he appeared in various other portraits as a confident leader, his physical body erect. Apparently to demonstrate that bin Laden had lost the war, the *Times* published contrasting stories about his ailing body. First, there was the story of his secret treatment for kidney disease (November 1, 2001), then the *Times* published a photographic triptych to visually demonstrate the gradual physical decline of bin Laden, also expressed in the headline "Going downhill: the decline of a tyrant" (December 26, 2001).

Pictures from the preinvasion period had invariably depicted bin Laden as a strong and confident leader. *Le Figaro* (September 12, 2001) had published a photograph of bin Laden from the waist up, in a camouflage jacket, sitting down, and a Kalashnikov leaning against a wall behind him. The picture showed all the attributes that bin Laden used to present himself with on the global stage. His camouflage jacket and Kalashnikov suggested war, his white turban and scarf Arab culture and Pashtun allegiance, and his beard Islamic tradition. In contrast to Bush, who was depicted on the opposite page with his eyes cast downward, bin Laden looked reassuringly at ease. The caption reinforced the framing of bin Laden as the confident Arab fighter with a quote from a bin Laden video released in June 2001: "Osama bin Laden: 'The walls of oppression and humiliation cannot be demolished but in a rain of bullets'" (September 12, 2001). On September 21, 2001, the *Times* published a youthful, confident and relaxed bin Laden. The caption reminded the viewers of his evil intentions and reputation: "Osama bin Laden during the Afghan–Soviet conflict of the 1980s. He has vowed to send Americans home in body bags." The same photograph was published in other newspapers too. The caption in *Le Monde* (September 25, 2001) drew attention to his smile, perhaps to suggest a sardonic nature.

A caption accompanying another portrait of bin Laden described him in *Le Figaro* as "the inspirator and perhaps even instigator" (September 13, 2001) of the 9/11 attacks. The map of Afghanistan in front of bin Laden in this picture suggested that he was either in control of the country or seeking it. The picture contrasted sharply with the photo on the September 14, 2001 front cover of *Le Figaro*, which depicted Bush near the damaged Pentagon building. In the photograph, it looks as if Bush is running away from the ruined Pentagon building.

Several photo-texts stressed bin Laden's global status. A photo-text in the *Times* (October 10, 2001) focused on his properties around the world, and indicated the U.S. origins of his camouflage jacket and watch. The attributes made bin Laden not just one of *them* but part of *us* too. He was portrayed as the center of a network that took not only global but—hyperbolically—"galactic" proportions in the *Times* (September 21, 2001).

Perhaps the most important characteristic of bin Laden in the press has been his appearance as an image within the image. At the start of the conflict, virtually all newspapers discussed here published a photograph of the American dead or alive poster with bin Laden's face, and bin Laden entered the news as an image within an image on various other occasions too. There were photographs with images of bin Laden on book covers, demonstrators' placards, and television screens; photographs depicted the face of bin Laden as a carnival mask used in Mexico, on mobile phone screens, and behind a bull's eye; and the *Times* (October 8, 2001) published a picture of a bin Laden body double. These copies of the "real" bin Laden created an additional layer of distance and a hyper-real effect: bin Laden as a mere image or symbol, like the flag. In a doctored image on a demonstration poster with TV character Bert from the children's series *Sesame Street*, bin Laden became an image among many others circulating the world.[23] *Die Zeit* (November 29, 2001) published a collage that repeated the same portrait of bin Laden more than 300 times, gradually decreasing its size to less than 2 $cm^2$, symbolically indicating bin Laden's gradual loss of power and relevancy. As noted earlier, such representations can be seen as part of a larger narrative construction meant to provide the intervention with a clear beginning, middle, and end.

After the fall of Kabul and Kandahar, photographs of Hamid Karzai began to enter the press. As Fowler (2007) noted, the British press changed its framing of Karzai considerably once it became known that he was backed by Washington to become the next president of Afghanistan. Whereas he was first described as "a tribal warlord," he later became "a military leader." When he became president, Karzai skillfully used several attributes to present himself as an acceptable representative of Afghanistan, both inside and outside the country. A Pashtun from the North, Karzai avoided typical Pashtun garments as well as the Western clothes that he used to wear earlier. Instead, he chose to wear the *qaraqul*, a hat that signals status but not ethnicity, and the *chapan*, a traditional garment worn over the shoulders across Afghanistan and Uzbekistan (Vogelsang, 2003). The press took over this frame and focused on several of these attributes, reframing Karzai as the elegant statesman and a true alternative to "primitive" Afghanistan.

One Afghan leader, who attracted much attention and admiration from the West, was the deceased Ahmad Shah Masud. His untimely death, 2 days before 9/11, and his funeral generated a great deal of interest in all the western European newspapers. His life was recounted in near-hagiographic terms, and he was frequently identified in an affectionate way by his nicknames, the "Lion of Panshjir" and "the Eagle." Photographs of his funeral depicted grieving individuals and crowds (including an unusual photo of weeping *mujahedin*), indicating widespread support for Masud. The press portrayed Masud as a peacemaker and as the brave fighter who resisted the Soviet army as well as the Taliban. That Masud was a Tajik and did not belong to the dominant Pashtun population made him an even more attractive model for the West. He represented an attractive third way, between Soviet communism and religious fundamentalism, with diverse ethnic groups in the country supporting him. The emphasis on the role of Masud as a peacemaker suggested that peace depended on individuals and that, with Masud's death, war had become inevitable. *Le Monde*, for example, published a photograph of Masud in plain clothes, symbolically looking away from a tank behind him, above the headline "When Masud dreamed of peace" (September 18, 2001). Apparently, with the death of Masud, the dream of peace had come to an end.

## HOME FRONTS IN A GLOBALIZED WORLD

Early news images in the period between 9/11 and the first bombardments on Afghanistan radically simplified the world by splitting it into two home fronts. A few photographs depicted daily life activities and activities supporting the war effort. The *Times* published a photograph of civilians who donated blood, with a caption that emphasized the extent of the support of war by speaking of "a rush of volunteers" (September 22, 2001). But most newspapers focused on the demonstrations. Civilian demonstrators in the West were framed differently than demonstrators elsewhere in the world. The use of contrasting frames created a dichotomous view of the world that reinforced the stereotypical image of a violent Islam and a peaceful West, and an interpretation of the world in terms of a clash of civilizations. Western demonstrators were portrayed as "anti-war," and occasionally as "anti-terrorism" or "anti-Taliban" (e.g., in *De Morgen*, September 25, 2001). *Le Monde*, for example, published a photograph of a man demonstrating in Washington with a placard with "NY♥PEACE" (October 2, 2001). The caption referred to the demonstrators and their commitment to avoid "the destruction of other innocent lives." Outside the West, demonstrators were portrayed as "Muslim," "anti-American," "anti-U.K.," and supportive of bin Laden or of the Taliban. Captions regularly identified the conflict as one between the Taliban and the United States, but further often generalized the Taliban to Islam. Pakistani demonstrators and Muslims were represented textually and visually as "religious warriors," suggesting a *clash of civilizations* frame. Two opposite rhetorical moves were regularly made: one in which Western identity was narrowed down to "American" (or sometimes "British"), and one in which Afghan, Pakistani, and Eastern identities were extended to "pro-Taliban" and "Muslim." The latter extension created an image of a large and united Islamic home front.

Protest pictures were either long distance shots emphasizing the size of a crowd, sometimes with an individual demonstrator in the foreground, or medium to close shots of a single demonstrator. Individual demonstrators with anti-American or antiwar slogans supported a dichotomous East–West framing. The *Times* published a photo of a young boy outside the Pakistan High Commission in London with the placard "USA at war with Islam" (September 19, 2001). Another photograph depicted a protestor in Jakarta, Indonesia, in close-up with a headscarf with the lettering "Go to hell USA" (*Times* and *Guardian*, October 11, 2001), and the *Times* also published a photograph of women demonstrators in close-up, carrying toy rifles and a placard with the slogan "Crush America." The picture was captioned "Pro-Taliban women demonstrating in Lahore yesterday: the rifles they are brandishing are toy, but their hatred of America is real" (September 20, 2001). *De Standaard* used the same picture, with a caption that described the conflict in the same bipolar way, as one between the United States and the Taliban regime but it added the religious qualifier "Muslim" to describe the women demonstrators: "Muslim women, who support the Afghan Taliban, are carrying toy rifles and anti-American slogans during a demonstration in Lahore (Pakistan)" (September 20, 2001). The same photograph—now in its uncropped version—was used in *Le Monde* and revealed a demonstrator who put the Quran up in the air. The caption in *Le Monde* drew attention to the religious attribute and the toy arms and suggested a clash between the United States and Islam: "The Quran in one hand, fire arms (in plastic) in the other, Pakistani activists demonstrated, Wednesday 19 September, in the streets of Lahore while chanting anti-American slogans" (September 21, 2001).

As journalists entered Pakistan in anticipation of the war, the number of photographs of street demonstrations in that country grew. Photographs of demonstrations in other Islamic countries were added, which contributed strongly to the *clash of civilizations* frame that was reinforced in headlines, captions, streamers, and quotes such as this one, attributed to Mullah Muhammad Hassan Akhond, deputy chairman of the Taliban Council of Ministers: "The United States, the Jews, the Christians and all those who support them, have as objective the destruction of the Islamic order" (*Le Figaro*, September 19, 2001). The accompanying photograph, taken in Quetta, Pakistan, depicted a clergyman putting up posters to gather support for a jihad. Depicting the sale of bin Laden paraphernalia was one of several ways to visualize support for bin Laden. Others were depicting the recruitment of young boys for jihad, focusing on the global reach of Islamism (including Islamism at home), emphasizing the size of the crowd of presumed jihadi protesters, and showing soldiers praying. Three headlines in a single issue of *Le Figaro* framed the impending conflict with a reference to jihad: "Taliban recruit for jihad," "Pakistani Islamists ready for jihad," and "Taliban prepare for holy war" (September 19, 2001). Photo-texts in other newspapers also indicated support of jihad. Among them a picture of a young cleric looking at a tee shirt that called bin Laden "a world hero," a photograph of a man selling a poster of bin Laden with the message "American will face another Allah's curse," and a young boy selling a turban with an image of bin Laden. The individuals were shown from the waist up behind their bin Laden gadgets. The photo of the cleric with the tee shirt was published in the *Times* and *De Standaard* on September 14, 2001, shortly after the attacks on New York and Washington. The caption in the *Times* accompanying this image associated Islam more overtly with violence, pointing to the rifle on the tee shirt: "A Muslim cleric in Peshawar, northern Pakistan, with T-shirts that proclaim Osama bin Laden the great hero of Islam. The shirts also depict a rifle and a map of Afghanistan." The shorter caption in *De Standaard* was "A Muslim cleric from Pakistan takes an interest in a T-shirt with a verse in honor of Osama bin Laden." Both photo-texts linked Islam to bin Laden without further qualification. Stressing widespread support for bin Laden, the *Times* also noted in the text of the article that the tee shirts "sold rapidly."

The Western home front was framed in several ways. Civilians in the West were depicted as fearful of terrorism, supportive of the war, against the war, oblivious to the war, or worried about war. One story in the *Times* on the reception of the war in what it called "small-town Britain" depicted elderly vacationers looking out at sea, apparently to the dangers that loomed across. The caption framed their lives in terms of fear of a nuclear holocaust: "All at sea: Holidaymakers and residents in Porlock, Somerset, fear what the future may bring. The town is only 30 miles from Hinckely Point nuclear power station" (October 13, 2001). The headline historicized the intervention in Afghanistan by quoting the vacationers: "At least we knew what we had to do to beat Hitler." The adjoining picture, which showed part of the small town of Hannibal, Missouri, suggested solidarity with the United States and hence a united Western front. Another article in the *Times* framed the American homeland in terms of fear too: "Fortress America learns to live under a terrorist siege." The constant stream of stories of anthrax attacks in all the newspapers further contributed to a double-sided discourse in the press: that of the war abroad and of the threat at home. The latter legitimated the war because it stressed what life might be under terrorist siege.

The most striking feature of home-front coverage in the *Times* was the virtually complete absence of photographs of war protesters. Instead, at the start of the intervention, the newspaper published photographs that demonstrated how sections of the population supported the war. There was an image of leaders of various religions gathered at a press event to illustrate widespread support for Blair, and two images depicted members of the public donating blood, with captions emphasizing the scale of the donations. They were qualified as "up 70 per cent" and referred to as a "rush of volunteers" (September 22, 2001). These images evoked the idea of a total war according to which war is not a battle between armies but between entire nations, including their civilian population.

The *Times* hardly ever described the Western audience as adverse to the war but it did occasionally represent the Western public as oblivious to the war. The article accompanying the photographs of the blood donations somewhat contradictorally stated in its headline that "Britain stays at home and watches TV" (September 22, 2001). *De Morgen* framed the public in an ambivalent way too. On the one hand, it evoked the image of an oblivious public in the headline "And the Belgian, he continued his travels" (September 18, 2001). On the other hand, the accompanying photograph of two tourists on a camel near a pyramid rejected the frame of the careless tourist and linked Islam to potential threat: "The Belgian tourist considers the Islamic country Egypt as a 'suspicious' zone." The press regularly represented Muslims outside Afghanistan as suspicious. Photographs depicted Muslims watching Al-Jazeera, implicating them in the media war. Associating British Muslims with Muslims in the Middle East, the *Times* carried a picture with the caption "Arabs in London watch war updates on al-Jazeera" (October 10, 2001). The article reported: "Tony Blair's televised message to the Middle East that the West was not at war with Islam appeared to have little immediate impact on a sceptical region." This suggested that the *clash of civilizations* frame originated in the East rather than the West.

*De Morgen* associated Western Muslims with holy war in an article entitled "Holy war in the West continues" (January 9, 2001). The article was accompanied by a photo of children looking out of the window of a mosque in Brixton, suggesting that the children were implicated in the "holy war." Another photograph in *De Morgen* (December 13, 2001) depicted German police during a search operation against 21 Muslim organizations. Muslims in Pakistan frequently were portrayed as extremists and supporters of bin Laden; Muslims "at home" often were depicted with ambivalent and negative connotations, especially in the *Times*.

Total war requires a united home front, and within the framework of a total war, civilians who do not participate in the war effort are typically perceived negatively and may even be treated as suspect. Total war, then, implies the idea of an "enemy within": Some civilians within the country may not be supportive of the war effort of their nation, and may even be adverse to it. Muslims in the West were presented as such potential "enemies within." A photograph of Prince Charles. who met with British Muslims, accompanied a *Times* article entitled "Young Muslims ready for 'holy war'" (September 29, 2001). The caption explained that the prince had visited a mosque in Hackney, London, "to demonstrate his support for British Muslims." Title and caption created a frame that associated Muslims with holy war, and simultaneously proposed that the nation, embodied by Prince Charles, supported them nevertheless. *Le Figaro* framed French Muslims in a similar way. They were represented as divided "between a quasi-unanimous condemnation of terrorism and solidarity with their brothers threatened by the Western coalition" (September 21, 2001). The frame, which put Muslims in an in-between position, reappeared in a story accompanied by a photograph of American Muslims praying in Dearborn. The story was headlined: "The dilemma of Arab-Americans between Quran and Star spangled banner" (*Le Figaro*, November 21, 2001). Several phototexts suggested allegiance of Western Muslims to the cause of the Taliban. A photo of a Taliban graveyard was used in the *Times* to illustrate an article entitled "Corner of foreign field that is England" (November 24, 2001). The caption pointed to "two British Pakistanis who volunteered to serve alongside Mullah Omar's men and who lie buried with Afghan and Arab fighters." A photograph accompanying an article about the "hunt for militants" was captioned "Protests in Karachi at the U.S. and British air strikes in Afghanistan. Neither MI5 nor MI6 realized that young volunteers from British mosques were signing up for weapons training in Afghanistan and Pakistan" (November 24, 2001).

Other stories were negative about British Muslims too. The *Times* ran the story of Anwar Khan, a British Muslim imprisoned by the Northern Alliance and accused of fighting for the Taliban. The caption reinforced the stereotypical representation of Muslims as almost "naturally" violent by quoting Khan: "Everyone has a hobby, like a football player ... I liked to play with bullets" (October, 2, 2001). In another issue, a picture of Ahmed Omar Sheikh, in jail in India, was contrasted with pictures of his school years in East London. Suggesting that one can never be sure of someone's identity, the article's streamer said: "There was no clue that he would become one of world's most wanted terrorists" (October 4, 2001).

Two calls were made upon British Muslims: They needed to confirm their allegiance to the British nation, and they had to improve their image. The *Times* used figures of authority to voice these calls. On October 4, 2001, shortly before the first bombs hit Afghanistan, the newspaper put a large portrait of Thatcher on its front page, below the headline "Thatcher speaks out on terror." A quote from the former prime minister below the headline said: "I have not heard enough condemnation from Muslim priests," and the portrait was captioned: "Baroness Thatcher says that Muslim clerics should stand up and say terrorism 'is not the way of Islam'." The newspaper raised doubts about the allegiance of British Muslims by echoing Bush's challenge that "you are either with us or against us." The frame that accused British Muslims of not doing enough reappeared in the *Times* in an article called "Blair strives to bring Muslims on message" (October 26, 2001). The article ended with a reference to Bernard Haykel of New York University "who has emphasized the need for Islamic clerics to condemn bin Laden." The newspaper published a photograph of Blair with the leader of the British Shia Muslims, a meeting reportedly set up to convince the Muslims to support the war effort (October 26, 2001). A few days later, the paper ran a picture of Queen Rania with her daughter Princess Salma, captioned "It was vital to show Islam as a religion of peace, she said" (October 30, 2001). The article's headline said: "Muslims 'must correct image'." A symbol of Arab modernity, Queen Rania reappeared in the *Times* on November 20, 2001—now on the front page. The press thus advocated a particular model for Muslims to adhere to.

## REFUGEES: THREATENED AND DANGEROUS

Western Muslims were not the only hybrid category in the press. Refugees also were portrayed in ambiguous and contradictory terms. Refugees featured prominently in the coverage of the period running up to the invasion of Afghanistan, when a large number of journalists had gathered in Pakistan near the Afghan border. Although images of refugees became less frequent when journalists began to focus on the movement of the Northern Alliance troops and on victims and prisoners of war, photographs of refugees reappeared regularly between October and December 2001.

In his study of the representation of Afghan refugees in television news between 9/11 and the U.S. attack on Afghanistan, Terence Wright observed that images of refugees in general are not "action packed" and that the stories about them "rarely reach a dramatic climax" but that instead refugees are more often treated by the media "like an insidious encroachment" (Wright, 2004, p. 98). Considering the case of the Afghan refugees, Wright concluded that "news media package the unfolding events to standard formulae" and that "the voice of the refugee remains at the end of a chain of 'framings': contextualized by the anchorperson, reporter, non-governmental organization (NGO) representative, and sometimes translator" (p. 108). My own analysis indicates that his observations are equally valid for news photography. Refugees rarely were shown in action. They were depicted as a crowd waiting to cross the border, or as a group squatted in front of their tents.[24] Refugees in these pictures were textually framed as passive, and often described in terms of a state of being ("separated from their family," "waiting") rather than doing.[25] Texts that did associate refugees with an action usually referred to refugees as "pouring into" or "flooding" an area, or collecting food or other types of aid. Refugees were generally depicted as either "threatened" ("hungry" or "persecuted") or "threatening" ("pro-Taliban"), both at the individual and collective level. This framing of refugees as vulnerable *and* dangerous was regulated by gender and family role: Women and children were almost invariably depicted as vulnerable, whereas adult males were depicted as vulnerable only *in their role of father* and otherwise as suspect or dangerous. With adult males, a long shot that creates distance often was preferred. When women, children, and fathers were portrayed, a close-up or medium shot kept the social distance relatively small and encouraged empathy.

The first photograph of an Afghan refugee after 9/11 appeared in the *Times* of September 17, 2001. The photograph was taken at Sangatte, near Calais, France.[26] The article focused on the "heightened tension between anti-American Afghans and other nationalities." The photograph accompanying the article personalized the refugee problem through a close-up of one of the refugees, Martin Reshad. The text described Afghans as dangerous and violent: "anti-American," "knife-carrying," cheering the 9/11 attacks, and threatening to attack the center's staff. Throughout the major part of the article, Afghan refugees were associated with the word "fighting." Only the final paragraphs mentioned that not all Afghan refugees in the camp shared anti-American feelings.

When the *Times* published its story of Sangette, other newspapers began to frame the conflict in Afghanistan through a focus on the refugee issue too. The frequent use of individual portraits of female and child refugees indicates a tendency toward feminization and infantilization of victimhood. The first photograph of a refugee in *De Standaard* depicted a woman in a makeshift tent in the refugee camp of Jalozai, Pakistan. The caption framed the situation as severe by pointing to the woman's age ("old"), the quality of her housing ("an improvised tent") and the separation from her family: "Under an improvised tent an old woman is waiting for the return of her family in the refugee camp of Jalozai near Peshawar, Pakistan" (September 17, 2001). The same photograph was reused later with a caption that refocused attention from the individual problem of the woman to that of many Afghans: "At the edge of the refugee camp of Jalozai near Peshawar, Pakistan, an Afghan woman is waiting for the return of her family. Everyday Afghan refugees continue to pour into Jalozai, even though Pakistan has officially closed the border with Afghanistan" (October 10, 2001).

A picture of female refugees on the move, with a girl looking into the lens of the photographer's camera, accompanied an article entitled "Fleeing families pay with their children's lives" in the *Times* (September 29, 2001). The caption contextualized the situation: "Afghan refugees arrive in Quetta, Pakistan. The drought and fear of U.S. air strikes convinced thousands to flee but many have died en route." The *Times* illustrated the refugees' plight by depicting what they had left to eat: "Roots, grass and white granular balls called raigak, an encrustation found on rocks, are all that is left to eat" (September 28, 2001). On the opposite page, the *Times* published a photograph of women and children on the run with the caption "A sick Afghan child cries in Pushta Nishin, Faryah in northern Afghanistan but all he can be offered to eat is weeds." A vertical panoramic shot depicted "women collecting water in Faryah." In both instances, the text framed the refugees as "struggling to survive." One picture of a refugee family was captioned "Maharan Ali with his family in Quetta. Persecuted for their faith by the Taliban, Hazaras have to pay hefty bribes to be smuggled out of Afghanistan" (*Times*, October 17, 2001), and a group of refugees depicted in another photograph was described as "hungry": "Pakistani guards keep a nervous watch yesterday near the border crossing at Chaman where thousands of hungry refugees have been prevented from entering" (*Times*, October 22, 2001). The caption, although qualifying the refugees as "hungry," evoked an atmosphere of danger by vaguely quantifying the refugees with the words "thousands" and describing the guards as "nervous."

In later reports, the *Times* also framed Afghan refugees in Pakistan as dangerous. A photograph of three men and a small group of boys was captioned "Afghan refugees in a settlement next to Quetta's airport have been ordered to leave because of fears that they could attempt to shoot down aircraft" (October 8, 2001). The newspaper turned its attention to refugees threatening journalists and to what it called "pro-Taliban refugees" (October 11, 2001). *De Morgen* did the same on October 17, 2001. On November 16, 2001, the *Times* published a photograph captioned "A Pakistani policeman strikes an Afghan refugee as tempers flare at the border of Chaman, yesterday. Hundreds of refugees and Taliban fighters have crossed into Pakistan," and later published a similarly structured photograph with the caption

"A Pakistani border policeman administers rough justice to a young Afghan accused of trying to seduce girls in a refugee camp" (December 1, 2001). The caption put the refugee ambivalently in the role of target (receiving the kick) and suspect actor (seducer of girls). Co-location of "refugees and Taliban fighters" promoted the transfer of attributes: Refugees became fighters, and vice versa. Common quantifications of refugees as being with "hundreds" or "thousands" evoked an atmosphere of threat.

In this context of the refugee problem, the international community was portrayed positively as a provider of aid. *Le Monde* (October 24, 2001) carried a photo of an Iranian doctor providing medical care in the Makhaki refugee camp, set up by the Iranian Red Cross. The *Times* published an article entitled "Woman on a lone mission of mercy" (November 9, 2001) with a photograph captioned "A British aid worker, Karen Davies, plays with children in the refugee camp she is running in Nowabad, northern Afghanistan, less than five miles from the front lines." Another photograph in the *Times* depicted "Daniela Charles, 4, helps Jemima Khan to launch a charity appeal to provide relief for refugee Afghani [*sic*] children" (October 24, 2001). The framing of aid by the press was generally consistent with that by U.S. public affairs in so far as women and children were typically framed as the recipients of aid, and Western women, including elite models such as Cherie Booth and Laura Bush, were cast in a traditional nursing role.

Some newspapers published photographs of refugees alongside articles that dealt with entirely different aspects of the conflict, unrelated to the refugee problem. It seems reasonable to assume that at least some of these photographs were primarily included as exotic points of attraction. Refugees appeared twice on the front page of *De Standaard*, probably for that reason, as their captions suggest. One image, published on the front page, was captioned "In buses, Afghans are trying to cross the border to reach the city of Quetta" (September 21, 2001) with the emphasis on "In buses" supporting the view that the ornamented buses were of greater interest than the refugees. The *Times* carried the same photograph on that day with the caption: "Afghan travel in buses on their way to the Pakistan city of Quetta, the capital of Beluchistan Province, after crossing the border at Chama." Apart from a problem of accuracy—in the *Times* the refugees had already entered Pakistan, in *De Standaard* they had not—both captions singled out the means of action: "in buses." Later on, *De Standaard* published another photograph of decorated buses, captioned "Never before did the Pakistani buses receive so much attention." *Le Figaro* (September 17, 2001) depicted refugees in buses too, but focused entirely on the refugees in the image. Evidence of the practice of drawing attention to exotic or photogenic features rather than structural problems also was found elsewhere. One aesthetically pleasing photograph showed "Fun far from home: Afghan children at a refugee camp in Islamabad enjoy a shower from a truck sprinkling water to damp down the dust" (*Times*, December 4, 2001). The use of a slower shutter speed blurred the motion of the water and enhanced the graphic quality of the picture, creating an aesthetically pleasing effect with the words "fun" and "enjoy" in the caption, apparently downplaying the gravity of the refugee problem. After the Taliban organized a tour for journalists to Spin Boldak, the *Times* published a photograph of a group of squatting refugees in front of their tents, a common view. These pictures were probably selected for their visual appeal too, as they were not accompanied by any discussion of the refugee issue.

One article dealing with the refugee issue in *De Standaard* (October 1, 2001), entitled "Central Asia awaits Afghan refugees in fear," and with the subhead "Fear of ethnic and political tensions," did not consider the refugee problem from the point of view of the refugees but focused on how "the refugee flood of an estimated 200.000 people" could be managed. The article was published with a photograph captioned as "Afghan refugees are leaving their country by all sorts of transport, for fear of an American attack but also for fear of the response of the Taliban. The Central Asian republics fear that an influx of refugees surpasses their means and—in particular—endangers their political stability." "Fear," which is repeated several times, is first related to the experience of the refugees but then to those who will be "flooded" by them. The Flemish newspaper regularly associated refugees with the "dangerous" transgression of borders. A photograph of a small group of young refugee children was captioned as "On the run. Afghan children in a refugee camp near the village of Kaga Bawden, in the north of Afghanistan, about 15 kilometers from the border with Tajikistan. That country expects that *tens of thousands of refugees will try to cross* the border" (September 24, 2001, italics added). The caption is structured in such a way that attention is shifted from the refugees to the neighboring countries, in this case Tajikistan. The qualification "tens of thousands" stresses the size of the perceived threat. The other Flemish elite newspaper, *De Morgen*, framed refugees as a threat too, in "Pakistan *fears flood* of another million Afghan refugees" (October 8, 2001, my italics). Framing refugees through their relation with the borderland setting also occurred in the *Times*' report entitled "Border closed to Afghan refugees." The accompanying photograph depicted a group of mostly male refugees having their papers checked at the border. Pictures of people crossing borders, and the patrolling and guarding of borders, were frequently published. Images of guarded borders appeared in all newspapers, sometimes with the border itself as the major point of attention. The border was depicted as an entry and exit point to the war zone. In the preinvasion period, images depicted refugees crossing Afghanistan's frontier to move into Pakistan. When the war started, newspapers began to focus on combatants from Pakistan entering Afghanistan. This became a more important issue when the Taliban began to use Pakistan as a base from which to operate. The photo-texts about border transgressions indicate that the western European press generally followed the view of its political elite that the war had to be confined to Afghanistan. They can be read as an expression of the fear that "the Afghan problem" would spill over into neighboring countries.

## THE CIVILIAN BODY AS BATTLEGROUND

The selective focus on attributes of the physical body has been used to create a stereotyped and gendered vision of the conflict in Afghanistan. Two body boundaries or forms of skin covering have played a key role in visually reporting the intervention in Afghanistan, and in defining the "liberation" of the country: the now well-known burqa, covering the face and hair of women, and the beard of Afghan men. Both are primary examples of the ideological use of synecdoche, that is, of parts representing the whole. They have been used in journalism as shorthand descriptions, generalizing the entire Afghan population into two social groups divided by gender. This generalized and gendered division of the population correlates to the official U.S. double discourse of militarism and humanitarianism, and has promoted a neocolonial view of the Afghan population as deeply traditional and overly determined by its own cultural codes. In 2001, the bearded Afghan warrior was a key element in the military discourse, whereas the burqa-clad woman featured prominently in the humanitarian discourse. The focus on these cultural attributes forced localization on the Afghan population and implied disconnection from the modern world. The removal of the beard and burqa after the fall of Kabul was supposed to mark Afghanistan's entry into the modern world. That this entry was construed around the erasure of Afghan cultural and religious attributes suggests the exclusion of alternative modernities.

Clothing and hair interface between the private and the public world, which makes both a target for those who seek to control people and define and alter the public world. Prior to the Taliban, the communists abolished the burqa to promote modernity, and the Taliban later reversed the situation and made it compulsory to promote their vision of Afghanistan.[27] In the West, the framing of the burqa as a symbol of oppression, and unveiling as an emblem of liberation, is related to a complex history of imagining the veil, itself an ambiguous term that covers a wide range of garments, customs and cultural practices.[28] The veil has been associated, among others, with sexual fantasies of male domination of Muslim women, virginity, and polygamy. Originally, the burqa signaled a high social status and urban lifestyle. Women whose husbands were employed by the government used it as a status marker (Noelle-Karimi, 2002). The burqa indicated that one did not have to work with one's hands, as this is difficult with a burqa on. It apparently acquired a new meaning in the refugee camps of Pakistan, where it became the common dress code for all women. The Taliban, who mostly grew up in these camps, considered it the only proper way to veil, and when they came to power they imposed it.

The dress has fascinated the press. Beyond the press, and in the wake of the 2001 intervention, articles, books, films, and laws have been made about the burqa. In 2001, caption writers almost never failed to indicate the burqa in photographs of women, using it as an attribute to collectivize them. A caption from *Le Figaro*, "Behind the mesh of their veil, Afghan women have barely the right to breathe" (September 18, 2001), sums up the common Western view that the burqa is a sign of cruel oppression. Captions, even when not overtly promoting the burqa as a symbol of the oppression of Afghan women, often drew attention to the garment. Singling out the dress made Afghan women look traditional, exotic, and trapped in their own past. A photograph of a woman going to hospital was captioned in *Die Welt* as "The Afghan woman Ziagol takes off her burqa when taking her small son for treatment to the hospital of Heart" (November 19, 2001). *De Standaard* headlined a story "Under cover of the burqa" (November 13, 2001). The *Times* published a photograph of a female refugee as "A burka-clad woman flees with her child across Afghanistan's border with Pakistan" (October 18, 2001). Photographs in the *Guardian* were similarly captioned: "A burka-clad refugee in one of two rooms her 25-member family shares near Peshawar, Pakistan, after fleeing fighting in Afghanistan" (October 18, 2001) and "An Afghan woman wearing a burka is passed by a Northern Alliance tank taking fighters to a new position near Jabal os Saraj, north of Kabul" (November 5, 2001).

Even before 9/11, the burqa was used as evidence of patriarchal domination and religious fundamentalism (Fahmy, 2004). Having defined the burqa in this way, the press looked for images of unveiling after the fall of Kabul and the liberation of Afghanistan. The *Guardian* published (November 15, 2001) a set of four photographs meant to illustrate Kabul's liberation. The largest of the set depicted an unveiled woman and was captioned as "A woman shows her face in public, Western dresses are displayed, locals examine pictures of film stars, and a boy chats with two girls as Kabul rediscovers freedom." *Le Monde* (November 22, 2001) published two photos of unveiled women, noting that a group of 200 had been able to walk in Kabul without their burqa. A Reuters picture of a young Afghan woman without a face-veil amidst other burqa-wearing women was circulated widely. The caption in *Le Figaro* stressed the unveiled face and made the symbolic intentions explicit: "An Afghan woman, with unveiled face. By itself, this photo could sum up the changes in Kabul after the retreat of the Taliban" (September 15, 2001). The same picture was captioned in *De Morgen* as "A brave young inhabitant of Kabul shows as one of the first her face in public during a food distribution in the city" (November 16, 2001). The same issue of *De Morgen* had another picture of unveiled women on its front page to symbolize the "liberation" of Afghanistan. The issue of *Le Figaro* carried another photograph of unveiled women too, captioned "After five years of Taliban yoke, the women of Kabul dare to walk down the streets with their face uncovered" (November 15, 2001). A photograph of an unveiled woman in the *Guardian* was captioned: "A 17-year old woman wearing only a headscarf is greeted by her neighbour in Kabul yesterday. However, despite the fall of the Taliban, many women are not confident enough to dispense with the head-to-toe burka" (November 16, 2001). On November 22, 2001, the *Times* published a photograph of "A widow who was happy to lift her veil during a visit to the UN World Food Programme bakery in Kabul."

Women who appeared unveiled in the mass media attracted particular attention. The *Times* published a photograph of Maneja Rasuli, captioned "Maneja Rasuli appears unveiled on Television Faisahad, and, right, ready to go on air. But off-screen her face is hidden by the burka" (October 22, 2001). The headline of the article in the *Times* localized the news by referring to Anna Ford, the first female newsreader in the United Kingdom: "Anti-Taliban TV's Anna Ford hides her face in public." The text historicized the moment captured by the camera, suggesting that Afghanistan was now going through a similar historical phase of emancipation as the United Kingdom at the time Ford made her appearance as the first female newsreader on ITN. In accordance with the 19th-century evolutionary idea that developing nations are on the path to modernization shown to them by developed nations, the phototext does not envisage modernity as a lived and local experience but authorizes Rasuli's "modern" identity by referencing Western history.

*Le Figaro* published a photograph of the same woman, but shot from a different angle and with a caption that did not mark gender: "Kabul television has restarted its transmissions, five years after the network's closure had been ordered by the Taliban" (November 19, 2001). When *Le Figaro* reused the photograph on December 24, 2001, it quoted Reporters sans frontières to stress that a female newsreader is not enough evidence of openness, and by early January 2002, the same newspaper had to conclude that "In Afghanistan, the burqa survives the Taliban" (January 3, 2001).

Members of the Revolutionary Association of the Women of Afghanistan used the burqa as a tool of subversion, to conduct clandestine operations during the Taliban regime (Brodsky, 2003). Without their burqa, they would not have been able to film and photograph the atrocities of the Taliban. In 2001, reporters like Yvonne Ridley of the *Sunday Express* and Michel Peyrard of *Paris Match* used the burqa in a similar subversive way. Dressed in a burqa, they crossed the Afghan border without a visa but were captured by the Taliban. BBC's John Simpson, who later famously reported that the BBC had liberated Kabul, also managed to enter the country in a burqa, and did not get caught. Suicide bombers have used the same tactic and have carried out attacks while dressed in burqa. These examples suggest that the burqa provides those who wear it with the power of the voyeur: *to see without being seen*. The burqa, then, is about the relationship between vision and identity, the visible and the invisible, the permissible and the impermissible.

Eyes and seeing were also the focus of a portrait of Sima Wall, the head of former king Zahir Shah's delegation at the Bonn conference, published on the front page of the *Guardian* of December 6, 2001; and the *Times* published a close-up of Cherie Booth imitating the veiled view of women in burqa on its front page (Fig. 3.5). The focus on the eyes of Booth and Wall suggests that vision is indeed central to understanding Western conceptions of Afghan women and the burqa. The captions seem to affirm such an interpretation. The photograph of Cherie Booth was captioned "Through Afghan eyes: Cherie Blair at her meeting with Afghan women and Cabinet ministers at 10 Downing Street yesterday," that of Wall entitled "The eyes have it." Mixing a Western face with an imagined burqa, the photograph of Booth does not merely represent an affiliative gesture. Booth's hybrid act can be related to a long history of Western women wearing the Oriental veil. At the time of the intervention, several popular culture celebrities, including Christina Aguilera, Lil'Kim, Feiticeira, and Britney Spears, put on veil-like dresses and continued this tradition. Their behavior cannot be separated from Occidental fantasies about Oriental sexuality, and one source of the desire to unveil Afghan women may be that the garment reverses the dominant Western cultural position in which women are traditionally the objects of the male gaze.[29]

The desire to lift the veil may not reflect an assumed moral duty to liberate women from oppression but rather the reverse may actually be true: Unveiling can be thought of as the expression of a desire to bring women under the male gaze and thus restore the naturalized Western order in which women are the object of vision, and men the surveyors. A Benetton advertisement seemed to do exactly this when it unveiled an Afghan woman by juxtaposing images of her with and without a veil (Mollison, 2002).[30]

Although unveiling and revealing the female face and hair in public as a symbol of liberation dominated the Western press around the time of the fall of Kabul, in the popular imagination, it was not just the face that was rendered visible but the entire body. Images of semi-naked

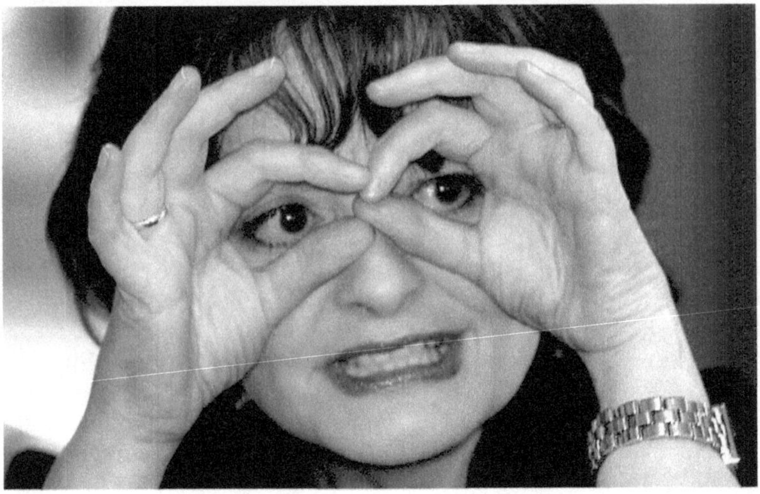

**Fig. 3.5 Cherie Booth. © Ian Waldie/Belga (2001)**
Published on the front page of the *Times*, November 20, 2001.

women in burqa-like clothing circulated the Internet and they appeared frequently in a variety of nondocumentary images too (Rantanen, 2005). This seems further evidence that a desire for the restoration of male voyeurism is an important element of the Western obsession with the burqa. If veiling is in opposition to the dominant Western gendered gaze, then unveiling restores women as objects of the male gaze. What has been presented as women's liberation actually contains a strong element of sexism: The female body is treated as a territory that has to be conquered through unveiling it.[31] That press representations of female agency in Afghanistan have largely been restricted to acts of unveiling supports this view.

There are two other aspects worth noting in this context. First, the idea that the West can create the opportunities to lift the veil of Afghan women and make them see the world resembles the cultural colonialist position according to which the colonizer is the seer who reveals the world to the indigenous population. Second, with veiling framed as a symbol of Islam, unveiling at the time of the fall of Kabul stood for a victory of modernity over an assumed a-modern Islam.

While the burqa has been widely used to define female Afghan identity, beards and headgear have been central to visual definitions of male Afghan identity.[32] Beards featured prominently in cartoons in *Le Figaro* and *De Morgen*, but there also were photographs. The *Times* published a picture captioned "A man had his beard cut on a street in Kabul after the fall of the city. Shaving had been forbidden by the

Taliban" (November 14, 2001). The phrase indicates that the male body was a site of struggle, but it appeared in the news as such only after the fall of Kabul, when photographers were able to picture the shaving of beards. On November 15, 2001, the *Times* headlined "Busy barbers trim Kabul men's devotion," and suggested that liberation would not be complete until veils and beards were totally removed. The caption accompanying the article's photograph said: "Even though there is a feeling of liberation in Kabul, and women are going out alone, only a few are doing so without the burka. Most men are having their beards trimmed rather than going clean-shaven." Because veils and beards were closely associated with Islam—beards were described as "men's devotion"—the implication of the text seems to be that Islam and a state of liberation are incompatible, and that the liberation of Kabul was not complete until the men gave up their devotion entirely.

Anthony Synnott (1987) defined hair as both personal and public, voluntary and malleable "and therefore singularly apt to symbolize both differentiations between, and changes in, individual and group identities" (p. 381). In his sociology of hair, he called the beard "a focal point and prime symbol of new identities and new ideologies" (p. 400). In accordance with these observations, the cutting of hair and the shaving of beards at the time of the liberation of Kabul symbolized a change in identity and an exercise of social control. At a pragmatic and local level, the trimming of beards indicated allegiance to the victor, but, symbolically, for the West, pictures of the shaving of beards also suggested the destruction of Afghan otherness and the conformation of Afghanistan to the modern world order. Shaving was seen, like unveiling, as a victory over Islam and as the successful disciplining of the Afghan body. If uncut long hair is a symbol of anarchy, freedom, and opposition—as among the hippies— the trimming of beards is not a sign of freedom but a restoration of order and control.

The beard also is a traditional symbol of male virility. More specifically, the Oriental beard is associated with Western sexual fantasies about Muslim polygamy. The shaving of Oriental beards, then, can be understood as an act of emasculation and effemination. The shaving of the skin indicates a change in the cultural and political landscape: The male Afghan body is symbolically subjugated to the victor, who, through the act of unveiling, also controls the female body.

More radical alterations of the body during wartime include mutilation and death. Dead civilians are indicative of total destruction and the end of community. Such images were rarely published. Images of wounded civilians were shown more often, and throughout the entire initial bombing campaign. Pictures of wounded children appeared in the Western European press from October 2001 onward, especially in the *Times*, even though the newspaper appeared at that time generally supportive of the war. As the war intensified, more photographs of wounded and civilian Afghans entered the press in 2001, but civilian casualties were later often reported in texts only—if reported at all.

## PHOTOGRAPHY AND INFO-GRAPHICS

Civilian casualties were a more important topic in cartoons than in photographs. The relationship between the cartoon and the photograph can be compared to the one between columns and news reports. The cartoon constitutes a more obviously construed space, one that need not bear an indexical relation to actual situations. Info-graphics are constructions too but generally presented as objective facts and therefore more closely related to photography. Info-graphic artists regularly use photographs as raw material for their designs, and they incorporate photographs in their work more directly. Photographs have been used as a background on which info-graphic elements are placed, and they have been integrated into info-graphics in various other ways. Unlike cartoons, info-graphics were rarely critical of the war in Afghanistan.

Many info-graphics of conflict focus on weaponry, soldiers, and other military aspects. They explain attributes of weapons and soldiers, indicate numbers of troops in geographical space, and use arrows and lines to frame war as a sequence of military operations. Info-graphics with a focus on the chronology of a conflict usually tend to suggest that war is a logical sequence of interrelated events, and those that combine maps with illustrations of weaponry and aircraft create an image of war as a technological operation on a geographical space. The frequent emphasis on technology and the abstraction of the landscape dehumanizes war and turns conflict into a technical operation devoid of human participation. Nations or troops are reduced to broad categories, such as "enemy," "friend," or "neutral." Info-graphics that "explain" war technologies tend to demonstrate, support, and celebrate the technological superiority of a nation. Many info-graphics invite the viewer to think as a soldier in a military and tactical way, and to vicariously partake in the action. For example, in 2001, elaborate visualizations of the possible whereabouts of bin Laden invited the viewer to become a part of what was called "the hunt."

One info-graphic strategy consists of laying portraits of leaders on top of geographical maps. For example, one info-graphic about the conflict in Afghanistan framed Blair as a global leader by visually summarizing what it called "Blair's 5,000-mile trip" from London to Moscow for a meeting with Russian President Putin, and from Moscow to Islamabad "via Chinese airspace to avoid Afghanistan" for a meeting with Pakistani President Musharraf. With the aid of photographic portraits, the intervention in Afghanistan was presented here in an *episod-*

*ic* way, as a conflict managed by world leaders. In other info-graphics, the focus shifted toward terrorists and "warlords" threatening the world.

Info-graphic maps were also used to display divisions. World maps featured in several info-graphics at the beginning of the conflict stressed the global dimension of the "war on terror." They often suggested a *clash of civilizations* frame, and invited viewers to take positions and locate themselves in the world. The *Times*, for example, used maps that divided the world into "the Islamic world against the intervention," and "the Western world supportive of the intervening coalition" (October 19, 2001). The *Guardian* and *Le Monde* tried to move beyond the military map on several occasions with info-graphics about Afghanistan's ethnic diversity and the movement of refugees. The ethnic maps invited viewers to think about Afghan heterogeneity as an impediment to state formation, and those of refugees were largely consistent with photo-texts representing them as "floods."

## INVISIBLE AFGHANISTAN

Gurevitch et al. (1991), in a study of international news on European television, concluded that international news events are framed "in ways that are compatible with the culture and the dominant ideology of societies they serve" in order to "render these events comprehensible, appealing and relevant to domestic audiences" (p. 206). This observation holds true for the newspapers discussed in this chapter, and is supported by an analysis of the portrayal of war leadership, combatants, and civilians. All newspapers discussed in this chapter deployed their own historical and cultural references to frame leaders, combatants, and civilians in ways that made them and the events in which they participated appear pertinent to their own domestic audiences. At a more abstract level, however, a number of *master* frames were shared across the western European press, with various levels of intensity. None of these frames were centrally related to Afghanistan or the Afghan people as agents of their own destiny. Afghanistan figured largely as an exotic background setting, and Afghan civilians mostly figured in subordinate and passive roles: as victims of the conflict or as beneficiaries of an aid program. Even when photographs depicted Afghans, captions frequently relegated attention away from them. The representation of Afghan refugees at the start of the intervention illustrates this point, with many captions stressing the consequences of the refugee problem for Afghanistan's neighboring countries and the threat to international political stability. When Afghans were depicted as actors, they often were portrayed negatively, as fighters, warlords, and drug dealers. The focus on Afghan combatants as wild and unruly, and later on the corruption of the government and the problem of the Afghan opium economy, created a frame in which the problems of Afghanistan were predominantly explained by reference to internal causes. The external origins of the conflict were explored in very few photo-texts.

Photo-texts that appeared sympathetic to Afghans were usually *episodic*, isolating the suffering individual in order to create an emotional effect. Suggesting compassion rather than engagement, it is difficult to see how such photo-texts could ever function as an aid for understanding the larger contexts in which the depicted individuals live. Captions, which are able to compensate for the singularity of images, were rarely used to provide the reader with a *thematic* frame. Instead of offering detailed thematic contextualization, captions tended to generalize the content of images. Commercial demands, space constraints, and routine practices encourage generalizations in the press at any time, but stereotypes made in the coverage of the early months of the intervention in Afghanistan indicate that many Western journalists were ill prepared to capture the complexities of the country. Later, the practice of embedding removed journalists from local civilian conditions, and promoted a military perspective with a focus on troops, tactics, and strategies. Often, press representations of Afghans were not much different from public affairs representations. In both, Afghans generally appeared as objects of the intervention even though press representations were occasionally more ambiguous as well. In ambiguous images, Afghans were allowed to perform "themselves" within "ourselves," as in the image of the American *buzkashi* player and the picture of Cherie Booth imagining the burqa.

Identities were framed in two major ways. Images of familial, military, and sometimes national and transnational homogeny suggested the need for social unity. This *unity* frame suggests that the theory of Bourdieu (1990), which holds that photography serves the social cohesion of the familial group, equally applies to larger-size imagined communities. It was thematically developed in scenes of soldiers in daily life, presented as "like us," in images of the (patriarchal) family, in photographs of patriotism, and in pictures of military, civilian, and political cooperation. With the exception of initial reporting in the *Times*, national unity was a relatively marginal preoccupation. Although an allied operation, the western European press framed the intervention in Afghanistan mainly as a U.S. affair, a view reinforced by the occasional image of American patriotism.

A more important *identity* frame for the press was *individual leadership*. Individual leadership was usually narrowed down to national leadership, and thus a *domestic* frame. When individual national leaders are presented as models for the nation, *domestic*, *leadership*, and *unity* frames converge. This was most evident in the *Times*, which domesticated the news about Afghanistan in the early stages with several portraits of Blair.

As noted earlier in the context of public affairs photography, individual leadership and unity are closely related to the *managerial* frame because a united nation and competent leadership usually are seen as necessary conditions for managing a war. In the press, the *managerial* frame was thematically developed in photographs that emphasized control through cooperation and surveillance and through technological leadership. With its emphasis on disciplined soldiers, supportive civilians, and sophisticated equipment, the *managerial* frame suggests that war can be a manageable and almost bloodless operation. As far as successful war management is seen as dependent on the support of a united nation, the *managerial* frame can be linked to the concept of total war. Other aspects of the frame relate it to the concept of just war. In just-war theory, when the principle of discrimination, according to which a combatant is not allowed to target noncombatants, and the rule of proportionality, which states that military action must lead to more positive than negative outcomes, are met, side effects are seen as regrettable but permissible.[33] The western European press did not reject all side effects as impermissible but did question the proportionality of ends and, to a lesser extent, the legitimate authority, one of the *jus ad bellum* principles. At the beginning of the intervention, western European newspapers criticized the scale of the bombing campaign and the harm done to Afghans civilians. Later, critique of the war in Afghanistan shifted toward the dangers faced by foreign combatants due to inadequate numbers of troops, improper equipment, and lack of decisive leadership.

The *humanitarian* frame is closely related to the *managerial* frame and legitimates war by referencing the *jus ad bellum* principle of just cause, loosely interpreted as a form of altruism (i.e., to help another nation).[34] The frame legitimates war by supporting the just-war principle of right intention: Military intervention is presented as an action motivated by the desire to bring freedom, progress, and prosperity to an oppressed populace. It suggests that war is not conducted to pursue national interests but to establish peace. The frame accepts the coexistence of humanitarianism and militarism, and legitimates the war as a means to nation-building. It counter-frames the postcolonial idea that holds that modern nation-states are largely founded on violence, often with the consent and complicity of their citizens.

The *destruction* frame focuses on these violent aspects. It is a negative *consequences* frame covering several themes: human suffering, material destruction, survival amidst destruction, and the breakdown of identities. In the press, images of human suffering reversed the portrayal of civilians as beneficiaries of the war and drew attention to the disruption of family life, to the victims of military action and other acts of violence, and to the precarious status of refugees. Close-up demand pictures of child victims created direct emotional appeal but offered no contextual explanation. Photo-texts of material destruction indicated the bad consequences of war for the local population but did not question the war option when the destruction was framed as "collatoral." Photographs of ruins and collapsed structures could have invited viewers to reflect on the ruination of culture and political structures, but they usually possessed a conventional dramatic quality that promoted quick consumption rather than contemplation and contextual understanding.

The breakdown of unified identities was mostly valued negatively and connoted as dangerous. Photographs of refugees, ID cards, borders, border guards, and airport security controls indicated a concern with identity, and a fear of the collapse of an imagined order based on stable identities and coherent communities.

The *ubiquity of war* has a temporal and psychosocial dimension. Photo-texts sequencing battles, wars, and war leaders suggested temporal ubiquity: History is war. Because newspapers typically select episodes, people, and locations from historical wars that are most relevant to their own national history, temporal ubiquity domesticates the news. The two world wars in particular remain an important frame of reference for western European journalists, and associations with these wars generally evoke the idea of a total and just war. The psychosocial ubiquity of war—the idea of the violent nature of mankind—was conveyed through images of the other as a "natural born killer" and thus relegated responsibility for war and violence to the other. It was suggested that, because of the violent other, war will always exist. Consistent with colonial imagery, Afghanistan was framed as a wild and untamed conflict zone, both attractive and repulsive to the Western mind. This brings to mind the Hobbesian idea of a permanent state of war. Occasionally, Westerners were sometimes represented as "natural born killers" too, but these were then seen as contaminated and contaminating hybrids. The *ubiquity of war* frame generally supported the "war on terror" discourse according to which the cause of Afghanistan's problems is internal (i.e., Islamic fundamentalism and terrorist training camps) and the solution is predominantly to be found externally (i.e., in the West). By predominantly attributing violence to the other, it was suggested that, unless the West introduced democracy, peace, and prosperity, war would continue endlessly. Only later, and following political elites, newspapers began to focus on internal *solutions:* the need for a proper Afghan army and police force, and for negotiations with some members of the Taliban.

The idea of ubiquitous warfare in a cultural sense is found in the *clash of civilizations* frame. The frame affirms differences and hostilities between primordial, homogenized groups, and suggests that war is caused by ideas, faith, or belief. The frame surfaced before the military intervention in October 2001 when most of the photographs of protests around the world were published. Western European newspapers usually did not frame the intervention in Afghanistan explicitly as a clash between the West and Islam, but narrowed it down to one between the United States and the United Kingdom and extreme Muslim fundamentalists.

The *hyper-reality* frame directs attention away from the actual event of the war to its representation, to how representations filter reality. The idea of a hyper-reality was conveyed through photo-texts of the media war, images that drew attention to their own status as an image, and images within an image. The frame relates to the concept of the *mise en abyme*, and to the idea that war can now only be accessed through images that are reproduced in perpetuity (Baudrillard, 1994, 1995).

Finally, the *absence of war* frame defines peace negatively, from a dichotomous perspective in relation to war. According to Rummel (1981), this is the popular Western definition of peace, as found in the Old Testament, ancient Greece, and international relations studies. Photo-texts, info-graphics, and cartoons about the conflict in Afghanistan rarely suggested or explored any other conceptualizations of peace. Info-graphics and cartoons did not go beyond any of the dominant news themes and frames found in photo-texts—even though cartoons were more critical of the war. To discover alternative perspectives, the next chapter turns to four photo-books about Afghanistan.

## NOTES

1. Readership is used as an alternative to circulation because, from an advertising point of view, the number of readers is considered to be more relevant than the number of newspaper copies. Readership figures generally are around three times higher than circulation figures. More information on circulation and/or readership figures can be found at www.ojd.com (French press), www.abc.org.uk (British press), www.ivw.eu (German press), and www.cim.be (Belgian press).
2. Some newspapers have become global news brands. The *Guardian Online* in particular has been quite successful internationally, and now appears to have twice as many readers outside the United Kingdom than it has inside the country (Levine, 2009).
3. Later, the American experience in Somalia (1993) was used to defend media control of the press. In Somalia, no press pool was set up and journalists were able to report freely on the operations in the country. It is popularly believed that images of starving Somalis led to the U.S. intervention in Somalia, and that pictures of a dead U.S. soldier dragged through the streets of Mogadishu also subsequently led to

the withdrawal of the United States (Sharkey, 1993). But images of starving Somalis became important only after the Bush administration had introduced it as a discursive theme and the withdrawal of U.S. troops was probably the result of positions taken by the Clinton administration earlier (Perlmutter, 1998).

4. Named after retired Army Maj. Gen. Winant Sidle, the panel was officially named Military-Media Relations Panel. The recommendations that it issued in August 1984 led to the creation of the Defense National Media Pool.
5. Dutch filmmaker Vik Franke, who spent 6 weeks with Dutch troops in Uruzgan, began shooting at Taliban when his camera batteries were finished (Müller, 2006).
6. De Zilveren Camera ("The Silver Camera") Award is organized every year by the Dutch association of photojournalists. Military photographer Sjoerd Hilckmann won the third prize in the category "foreign documentary photography" but his work was removed from the association's Web site when it turned out that he had altered one of his images digitally.
7. An Airborne Warning and Control System is an aircraft that provides real-time imagery of a territory to an operations center. It is used for surveillance, warning, control, and information-gathering purposes.
8. Publications under the Taliban regime included the weekly *Kabul Times*, the Pashto-language magazine *Nangarhar*, the Farsi-language newspapers *Hewad, Anis*, and *Sharia*, and the monthly *The Islamic Emirate* (Reporters sans frontières, 2000). In Northern Alliance controlled territory, TV Badakhshan broadcasted programs in Dari and Pashto from Feyzabad, the capital of the North Eastern province of Badakhshan.
9. They appeared to be less friendly toward Arab journalists. Al-Jazeera reporter Ali Al-Arab was sent back to Tajikistan by the Northern Alliance, perhaps because two men posing as Arab journalists had killed their military leader, Ahmad Shah Masud (Committee to Protect Journalists, 2001).
10. For reports and stories, see The International Press Institute at http://www.freemedia.at, Reporters sans frontières at http:// www.rsf.org/, and *American Journalism Review* at http:// www.ajr.org.
11. The so-called anthrax attacks started a few days after the September 11 attacks, on September 18, 2001. Envelopes with *B. anthracis* spores were mailed to media companies and government officials in the United States, killing five people and sickening several others.
12. Flynt challenged press restrictions on November 16, 2001. U.S. District Judge Paul Friedman denied his request on January 8, 2002, when the war was already over. See *Flynt v. Rumsfeld*, 180 F. Supp. 2d 174 (D.C. Cir. 2002) and *Flynt v. Rumsfeld*, 245 F. Supp. 2d 94 (D.C. Cir. 2003). Getty images commenced litigation on January 31, 2002. It contended that it had been excluded from the DoD media pools. See *Getty Images News Services, Corp. v. Department of Defense*, 193 F. Supp. 2d 112 (D.C. 2002).

13. The number is an overestimation of the actual death toll of 9/11. The official number of victims, excluding the hijackers, was first estimated at 2,792 by the City of New York. The number was later reduced by 40 to 2,752 (Hirschkorn, 2003).
14. One exception was a small photograph by James Nachtwey (VII), published below the fold inside a supplement of *Le Monde* (December 14, 2001). The photograph depicted the execution of a Taliban soldier by a Northern Alliance combatant on the streets of Kunduz. The photo is reminiscent of Eddie Adams' iconic "Saigon execution" but apparently did not attract much attention.
15. The *Times* published a photograph of a Northern Alliance soldier captioned "A Northern Alliance gunner yesterday. Many of the Taliban he is about to attack are former comrades who are expected to rejoin the alliance" (October 17, 2001). *Le Figaro* carried a picture of a Northern Alliance soldier embracing a Taliban combatant, captioned: "Symbol of this Afghanistan, where everything is surprising: a Northern Alliance soldier (*left*) embraces a Taliban who has just surrendered" (November 26, 2001).
16. Pasthunwali is the customary code of honor of the Pashtuns built around such general principles as courage, hospitality, gender boundary maintenance, and council (Dupree, 1973). Wahhabism is a form of orthodox Sunni Islam from Saudi Arabia and Qatar that adopted the Hanbali school of Islamic jurisprudence. Radical Wahhabism has been accused of financing and otherwise supporting terrorism (Armanios, 2003).
17. See, for example, "A combatant of the Northern Alliance looks at American F18 bombing the Taliban troops, north of Kabul" in *Le Figaro* (October 23, 2001), and "Afghan fighters observe the American bombardments on Al-Qaeda positions in the Tora Bora mountains" in *De Morgen* (December 8, 2001).
18. For a discussion of the persistence of colonial metaphors in the British coverage of Operation Enduring Freedom, see Fowler (2007).
19. For a critique of Azoy's use of *buzkashi* as a synecdoche, see Fowler (2007).
20. Gen. Pervez Musharraf tried to respond to demands from the Western alliance without alienating the Islamic parties in his country.
21. Under the German constitution, troops can only be deployed after acceptance by parliament.
22. On September 21, 2001, the *Times* published a photograph of the prime minister and his spouse Cherie Booth below the headline "Blair's finger on the trigger." The "finger on the trigger" image was repeated a few days later, on September 28, 2001 when the *Times* published a photograph of U.S. Air Force Maj. Gen. Larry Arnold. The caption said: "His finger on the trigger, U.S. Air Force Major-General Larry Arnold, Larry K., a veteran pilot, who is based in Florida."
23. The poster with Bert and Osama bin Laden was used in a demonstration in Bangladesh. The images apparently originated from a number of satirical Web sites called "Bert is Evil."

24. Pakistani guards did not allow Afghan refugees without a passport in the country.
25. See the *Times*, October 19, 2001: "Afghan children, right, in an earthen home at a refugee camp near Quetta. Below, a refugee in a camp in northern Afghanistan." See also the *Times*, October 31, 2001: "The refugee camp at Spin Boldak, three miles inside the Afghan border, which was included on the Taliban 'tour'. About 5,000 refugees are waiting to cross into Pakistan."
26. At the time of the report, approximately 50% of the population of the Sangatte center was Afghan. Following requests by the U.K. government, the French authorities closed the center in December 2002.
27. It remains unclear how successful the Taliban really were in enforcing the burqa throughout Afghanistan, and how many women experience the dress as compulsory (e.g., M. Rubin, 2000).
28. For a comprehensive account of the multiple meanings of the veil, see, for example, El Guindi (1999).
29. On gendered vision in photography, see J. Berger (1972). The positioning of women as objects of the male gaze in narrative cinema was explored in an influential essay by Mulvey (1989).
30 For a discussion of the advertisement, see Edwards (2007).
31. In her study of photos of Afghan women from the Associated Press archive, Fahmy (2004) observed that women were represented differently before and after the fall of the Taliban regime. After the fall, they were portrayed more on the equal angle, more from the front angle, and more in medium than long shots, which suggests more identification, more involvement, and a closer relationship. Afghan women after the fall were also depicted in less traditional and more interactive roles, according to Fahmy. Her observations were based on archive images, not actual published pictures.
32. Members of the Northern Alliance generally preferred short beards and the *pawkul*, a beret-like hat that became fashionable among the *mujahedin* resistance during Afghanistan's Marxist period between 1978 and 1992. The Taliban preferred longer beards and a turban. The turban distinguished them from the Tajiks who had come to power after 1992. When the Taliban came to power, they prohibited the use of the *pawkul*. When they were ousted from power, many men switched to the *pawkul* of the victor.
33. For a discussion of side effects in just-war theory, and the related doctrine of double effect, see Ceulemans (2007).
34. The just-war literature varies on what a just cause is. Narrow interpretations may limit it to national self-defense; broader interpretations may accept extreme oppression as a just cause for going to war (Rengger, 2002).

# 4

# The Photo Book

This chapter concludes the analysis of the relations between photography and foreign conflict with a look at a number of photo-books. The book is perhaps, as Parr and Badger (2004) see it, photography's ultimate destination. There certainly has been a close association between the two since the earliest days of photography. Even before halftone printing was invented in 1880, photo-books and portfolios were used to present images to the public. Later, when the Russian avant-garde revolutionized photography, the photo-book became an experimental and artistic project. Photographers began to use it to explore positions not generally presented by the popular media.

## PHOTOGRAPHY AND THE DOCUMENTARY TRADITION

The four books discussed in this chapter qualify, but in various ways and not in any straightforward way, as documentary. In photography, the documentary project started as a reaction against the pictorialist tradition that sought to emulate painting. It positioned itself as radically different through an emphasis on the straight recording of reality. Many documentary photographers also have framed their work against photojournalism. Rejecting the sensationalism and noninvolvement of photojournalism, documentary photographers have tried to reveal underlying social patterns to promote social change. In the United States, the photographic document as a form of social revelation is inextricably linked to the Farm Security Administration photographic project (1935-1944). In Europe, the Russian revolutionaries believed documentary photography could reveal the material base of the proletariat and raise class-consciousness, an idea developed further in the German *arbeiterfotografie* (*worker photography*). Alongside these traditions, construed around the concept of social transformation through visual revelation, there has always existed another and more detached version of documentary work with archival and aesthetic concerns. August Sander (1876-1964), with his huge taxonomic portrait project *People of the Twentieth Century*, is often quoted as a leading example. Another major figure of this dispassionate, understated approach is Walker Evans (1903-1975), one of the first documentary photographers on the Farm Security Administration project and now widely recognized as a photographic artist.[1] In the 1950s, a more subjective strand of documentary work emerged, most notably in *Life is Good and Good for You in New York* (Klein, 1956) and *The Americans* (Frank, 1959). The same subjective approach was later applied to the photography of conflict when Gilles Peress (1984) combined his photographic record of the Iran hostage crisis of 1979-1980 with a personal account in the form of telex messages of his life as a freelance photographer.

The personal, socially concerned, and artistic orientations in documentary photography are not unproblematic divisions, and they cannot be separated from the political. As Burgin (1982b) and Tagg (1988, 2009) argued, all documentary work is political and positions the photographer, the viewer, and the depicted objects. The position of the photographer as a documentarian is never simply grounded in a transparent indexical relation of the camera to the real but the result of discourses and networks of power. Even Sander, despite his archival and aesthetic interests, ran into trouble with the Nazis because they believed that the German people he portrayed did not correspond to their Aryan ideal. His son was imprisoned, and the Government Bureau of Fine Arts ordered the destruction of his *Antlitz der Zeit* (*Face of the Time*).

## WAR AS QUEST

The books discussed in this chapter are related to the socially concerned, detached, and personal orientations in documentary work in various ways. In her introduction to *Afghanistan: The Road to Kabul*, Alison Morley (2002) of the International Center of Photography located Haviv's work primarily within the category of straight and concerned photography,[2] describing his photos as "just the light and the truth as Ron Haviv saw it that day." She made strong transformative claims, arguing that Haviv's photographs may provoke indignation and concern, and may generate aid and support. Haviv's photos are said to "bring compassion and understanding to the world," and it is suggested that we need his testimony "lest history eagerly repeats itself." The photographs are presented as moral warnings that may prevent the repetition of war.

*The Road to Kabul* contains 80 color images made during Haviv's 2001 journey with the Northern Alliance from the Afghan-Tajik border in Panshjir to Kabul. A longer documentary sequence of photographs combined with a text chronicling this journey, the book extends Haviv's regular photojournalistic work. The text by freelance journalist Ilana Ozernoy, who joined Haviv on the trip, is interspersed and breaks up the set of photographs into various parts. The captions are listed at the end of the book. Most of the images are laid out on two opposing pages, with a large white margin on one side. For the sake of rhythm, some spreads contain two images. The size of the book (12 x 10.6 x 0.8 inches), its hard cover, and its paper quality separate it in appearance from daily newspapers and press magazines in which Haviv's work is usually published. It can be read as a personal road diary, although its images mostly repeat the dominant news scenes, themes, and *master* frames discussed in Chapter 3. *The Road to Kabul* is an action-oriented *episodic* narrativization of the intervention in Afghanistan that largely follows the conventions of photojournalism and linear storytelling and the official U.S. double discourse of militarism and humanitarianism. There are pictures of Northern Alliance troops watching U.S. bombers in the sky, of smoke arising at the horizon, of veiled women, of children playing in the war zone, of Kabul residents in the cinema after the fall of their city, and of the death of a soldier. On one occasion, Ozernoy offers a glimpse into an exotic Afghanistan, when she speaks of "intricate gardens" and "the beauty of flowers in a barren landscape" (Haviv, 2002a, p. 31) but this theme is not developed further. The book's major symbol is the rough road to Kabul. It is called "a symbol for Afghanistan itself, of the brutality it has long faced and may yet endure again" (p. 7). The road in Haviv's work is a spatial symbol that equalizes war, violence, and Afghanistan ("war greets us with every curve of the craggy road," p. 8). This stands in opposition to the meaning attributed by USAID to that other road discussed earlier, the Kabul-Kandahar road (see Fig. 1.2).

The journey to Kabul gives the book its temporal structure. The story is organized as a traditional linear quest, with a plot, development, and conclusion. It is built around three major types of events or actions. It begins with the oppression by the Taliban, continues with fighting the Taliban, and then moves toward its resolution. As in traditional storytelling, the first photographs introduce the protagonists, the antagonists, the basic conflict, and the setting. Drawing attention to the territory, the first image depicts tank tracks in a deserted landscape. The first photograph of a human actor—and the second in the book—is of a woman presented as a tiny figure in the same desolate and barren landscape. The image connotes fragility, suffering, loneliness, and oppression, mainly through the posture of the woman in the image. The characteristics of the landscape, when transferred onto the depicted woman, reinforce this interpretation. The caption refers to the woman as well as the landscape, and frames the woman in the stereotypical way of the press, attributively, by describing her as "in a traditional burqa" (p. 158). A close-up of a Taliban decree that explains the laws governing women precedes this image. It is followed by an image of a man holding a PSYOP leaflet with a picture of a man whipping a woman. Taken together, these initial images suggest that the war is not about land, as the first picture seemed to indicate, but that it is about the oppression of women.

Although *The Road to Kabul* starts off by locating the intervention in Afghanistan within a *humanitarian* frame, the book is largely an account of the author's journey with the Northern Alliance and primarily a soldier's story. Evoking earlier colonial notions of the Afghan warrior, the combatants of the Northern Alliance are described as "natural born killers" who are insensitive to the threat of war. They "can only laugh in the face of incoming fire" as they are "born into war" and "dream of martyrdom with the conviction of a suicide bomber" (p. 31). They also are admired for their "ingenuity" and "their knowledge of the mountainous terrain canceling out the Soviets' technological superiority" (p. 7). The same combination of admiration and fascination surfaced in the press, and there also is similar attention to the blurring of boundaries. In a section called "Sometimes we exchange fire, but it's no big deal" (a title that frames violence as almost normal for Afghans), the text breaks down the dualistic view of the war as a fight between Northern Alliance troops and the Taliban, and explains how both depend on each other, and how people regularly change sides. This is not explained as a strategy to avoid further conflict but as its reverse, as a sign of Afghan whimsicality. At the end of the book, it is named as one reason why peace is a difficult goal: "America had cornered the bad guys, the terrorists, but it was impossible to imagine peace in a country where allies turned to enemies overnight, where warlords made government bedfellows, where corruption seeped into every crack of society" (p. 147). In accordance with photo-texts in the press, the responsibility for Afghanistan's problems is attributed to internal causes.

The middle section of the book contains photos of a dying Northern Alliance commander that are consistent with the book's focus on the brutality of war and the soldier's role in it. Following Robert Capa's dictum that "if your photos aren't good enough, you're not close enough," Haviv depicts, as a close witness, the death of a commander in a three-picture sequence. In the first, the commander collapses. In the second, he looks straight into the camera while dying. It is an unusual demand picture in which the commander is looking at the viewer in his moment of death. The third photograph shows how the commander is being carried away, and this finishes the episode. The viewer has witnessed death but without any knowledge of the commander's name, character, and history. The scenes that follow locate the commander's death within a struggle from oppression. Central to these images of liberation are the two gendered symbols of Afghanistan of the mainstream press: the burqa and the beard. In a section called "Now I have control of my beard!" men are shown having their beards trimmed, while gazing at photographs of Bollywood starlets. The text notes that women keep "their faces hidden underneath blue, floor-length robes," that "their small token of freedom" remains limited to "the absence of a male relative escort" (p. 85) but also that the burqa is an ambivalent instrument, "at once oppressive and protective" (p. 133). Locating Afghanistan between oppression and liberation, the book depicts women both veiled and unveiled. Some are portrayed in an educational setting to indicate liberation and change. In one image, two women in *chadri* are looking at a balloon seller. Their high-heeled shoes beneath their dresses seem to suggest a longing for Western fashion, and the balloons, forbidden under the Taliban, are an obvious symbol of freedom. The balloons, high heels, and the burqa frame Afghanistan symbolically as a transitional zone, between oppression and liberation.

Other images represent the horror of dead soldiers and mutilated civilian bodies. One photograph depicts "A funeral for a woman killed by an American bomb landed by mistake on a Northern Alliance village." Consistent with photojournalistic conventions, the woman remains a nameless, faceless victim. The caption objectifies and passivizes the agent of the killing ("by an American bomb"). "The wounded Afghan" in the book (pp. 34-35) has no name either and the circumstances in which he was wounded also are left out. The reader is thus offered an account in which particular victims and perpetrators are generalized into categories.

Some photographs depict houses with weapons, plans for attacks, forged visas, instruction books, and other materials that have been left behind. The text explains that here "we discover the real reason America joined the fight in Afghanistan" (p. 117). It is now suggested that the war is fought for security reasons, to combat international terrorism, not for the liberation of women.

The final images return to the promise of this liberation and the certainty of death and destruction in war. The penultimate photograph is a frontal color picture of five burqas hung on a wooden bar, "put out to dry" (p. 159) as the caption informs the reader. Beyond its literal meaning, the scene suggests that the burqa might be put away, and that the time of women's liberation has come to Afghanistan. The text confirms that "change is slow but quietly persistent" (p. 133). The next and final image of the book is of a flag on the grave "of a respected man" (p. 159). The war in Afghanistan is portrayed in these two final images as an uncertain liberation from what is represented mostly as an obvious oppression of women. It is framed as both constructive and destructive, and the framing of the conflict is gendered: Men are fighting for the liberation that will benefit women.

The *destruction* frame is found in images of human suffering, material damage, and oppression. The frame is partially countered by the *humanitarian* frame conveyed by images of liberation (unveiling, shaving, playing, going to the cinema) and traditional humanitarian and aid activities (schooling, food supplies). Another *master* frame from the press that dominates *The Road to Kabul* is that *war is ubiquitous.* The texts and photographs in the book portray Afghanistan as a perpetual war zone and attribute responsibility for the perpetuity of war to the Afghans themselves, to their corruption and violent disposition. The *ubiquity of war* frame is developed thematically in terms of the other as "a natural born killer" and history as war. Temporally, Afghanistan is framed in contradictory terms: untouched by time and touched by the history of war at the same time. Historical references are made to the time of the Soviet invasion but the role of the United States in pre-Taliban Afghanistan is eclipsed.

## DYSTOPIAN LANDSCAPES

Time and the ubiquity of war also are central elements in Norfolk's *Afghanistan: Chronotopia* (2002)[3] and Seawright's *Hidden* (2003).[4] Both are collections of skillfully construed aftermath landscapes that fit into the photographers' larger artistic oeuvre and that visualize the imprint of violence and history on space. Norfolk has focused on the militarization of the landscape in various parts of the world, and relationships between violence and the physical landscape also have been central in Seawright's work.

Both photographers distance themselves from the aesthetic familiarity of popular photojournalism and conventional resolution-oriented narration through choice of subject matter (landscapes rather than human beings), timing (photography *after* war), aesthetics (the use of color, fine detail, large format prints), presentation (book design), and textual choices (captions and introduction). *Chronotopia* and *Hidden* follow many conventions of the modern art book. Both are in oblong format, the photographs usually are presented on a single page, and their captions are relegated to the end of the book. Norfolk's work contains 47 plates, and is physically the largest book of the two (12.8 x 11.4 x 0.7 inches). It has a slightly more complex design than Seawright's book. In *Chronotopia*, single photographs occasionally are spread out on a double page, and the book has a section that folds open to display a single panoramic shot on four pages. *Hidden* is slightly smaller than *Chronotopia* (12.1 x 9.5 x 0.7 inches). The photographs in *Hidden* were commissioned by the Imperial War Museum in June 2002 and made during a stay with the UN and a U.K.-based de-mining NGO, the Hazardous Area Life-Support Organization. The book contains 28 images, one of which is a triptych. With the exception of this sequence, each image is laid out on a single page.

With their rich tones and warm colors, Norfolk's images in *Chronotopia* almost look like travel magazine photographs but content, paper quality, size, layout, and their production with a large-format camera position them more firmly as art. Norfolk describes the ruins as indicators of the suffering caused by war, a suffering that he believes is suppressed in the mainstream media. Ruins often have been used in painting, photography, and literature to warn against the destruction of culture and civilization, and Norfolk locates himself in this tradition in the introduction of his book. In photography, Richard Peter's *View from the Dresden City Hall Tower Toward the South*, is probably the most iconic example. Peter's image of Dresden after World War II offers a bird's eye view of the city's ruins over the shoulder of a bending female stone figure on the city hall tower. The figure called *Bonitas* (*Goodness*), a 1910 statue by Peter Pöppelmann, symbolically represents the order of the city government. It points its hand— apparently in accusation—toward the chaos of the ruined city beneath (Koetzle, 2002). Norfolk's *Chronotopia* possesses the same moral quality. The ruined landscape in the book is an obvious metaphor for Afghanistan, its people, and its history of war. The ultimate reference point in *Chronotopia* remains civilization: The ruins in the images are of a government building, a teahouse, an apartment building, a bazaar district, a swimming pool, and a royal mausoleum. These offer the viewer a glimpse of what once constituted a high cultural order. In his captions, Norfolk explicitly draws attention to the civilized functions of these buildings prior to their destruction. The mass media have generally worked from an a-historical perspective, associating Afghanistan with the Stone Age; several pictures in *Chronotopia* suggest a grand cultural history. The moral condemnation of its destruction is reinforced in the introductory text, which compares the photographs with the style and subject matter of French Baroque landscape painter Claude Lorraine (1602-1682) and German Romantic painter Caspar David Friedrich (1774-1840). The reference to these painters is part of a larger textual explanation meant to inscribe Norfolk's work into the tradition of the 18th-century sublime. The sublime, often represented by an unspoiled natural setting that creates a sense of human powerlessness, is both overwhelmingly beautiful and horrific. It is a form of "delightful terror" (Burke, 1998, p. 67).[5] In accordance with this notion, Norfolk's work is intended to offer viewers a pleasurable aesthetic experience and to confront them with the consequences of a repulsive history of war at the same time. The picturesque style is used as a strategy to encourage contemplation of the image and as an antidote against the quick consumption of photography.

Contra one of the ideas of the mainstream press and public affairs, that war is manageable with the right tools and leadership, *Chronotopia* frames war as *ubiquitous* and *destructive*. The work conveys a deterministic, pessimistic, and dystopian vision. The prospect of peace, freedom, and prosperity barely exists in the photographic universe of Norfolk, and if it does seem to exist, it survives *despite* the war. There are few human individuals in his book, but those that do occur function by virtue of their props as figures of hope and liberation. A man with balloons, a man with a bird, grazing animals, and a herder and his sheep all seem to suggest that a return to normal life remains possible. These isolated characters introduce an element of life and movement in an otherwise ruined land.

The photographs in *Chronotopia* are presented as evidence of places that have been revisited by war time and again. Instead of the idea of *war as history*, found in the mass media and in Haviv's *The Road to Kabul*, which projects the intervention in Afghanistan in linear time, Norfolk's photos evoke a multitemporal frame referred to in the foreword and book title as Mikhail Bakhtin's "chronotopia." War is treated as an accumulative phenomenon, particularly in those images that depict rusting airplanes, vehicles and equipment, and ammunition from former wars. Afghanistan is seen as a perpetual war *target*, not a zone of inherently violent inhabitants, a common view in the press. Much of the war debris in the pictures is actually from the Soviet period, and demonstrates that Afghanistan's history of violence is one of *outside* forces.

Seawright's *Hidden* evokes the *ubiquity of war* frame too but in a different way. The landscapes in *Hidden* have a very different quality than those in *Chronotopia*. Instead of the golden light of Norfolk's photographs, Seawright's images have cold monochromatic tones. Instead of the accumulation of ruins and debris, the aftermath landscape of Seawright is empty. Whereas Norfolk's *Chronotopia* draws attention to the directly visible impact of war on the landscape, *Hidden* presents the viewer with the barely visible remains of warfare. Like Norfolk, Seawright's images are meant to counter photojournalistic representations. Stereotypical representations of refugees are replaced by images of empty refugee tents. The book does not confront the viewer with the ruined skyline, but with the abandoned interior. Bomb explosions, wounded civilians, and dead soldiers are replaced by images of hidden landmines. There is hardly a trace of the violence that has occurred, and no trace of those that have witnessed it.

The landmines indicate past warfare and present dangers. The empty refugee tents and abandoned interiors suggest the complete erasure of even the smallest private existence. All the cultural signs that define a landscape are absent. War seems to have left virtually nothing behind. Only minor traces of human existence remain: small heaps of sand, check marks on walls, nails, and a map. A few captions—"Minefield," "Camp boundary," "Map," and "Perimeter"—indicate that these are war landscapes. The viewer scrutinizes the images and the captions only to find them empty of explanations, patterns, and causes. *Hidden* frames war *as a complete void*, empty of any meaning. There is no chronology, no reference point. All topographic and narrative information is excluded. The viewer finds himself in an unfamiliar, empty and incomprehensible desolate space. In *Mounds* (Fig. 4.1), neither the photograph nor the caption reveals that the heaps of sand are the result of de-mining activities. This creates an open interpretive space for viewers, in which they may think of the heaps as graveyards and of war as equal to death.

**Fig. 4.1. "Mounds, Afghanistan 2002."**
**© Paul Seawright. Courtesy Kerlin Gallery.**

Because human actors are physically absent, action and human suffering are implied only. Those who laid the landmines, those who removed them, and those who may be killed by the mines that remain, are all kept out of sight. The camels in one landscape are the only sign that life is not entirely wiped out. This photograph is reminiscent of the colonial landscape, empty of human beings, seducing viewers to enter it to become part of the imperialist project (Mitchell, 1994a). But Seawright's mined landscapes subvert the imperialist temptation: His is not the colonial virgin territory waiting to be exploited, but a raped and hostile land. His work signals the annihilation of nearly everything—the private as well as the public.

The a-spectacular and seemingly dispassionate approach of Seawright can be traced back to the highly influential landscape photography exhibit *New Topographics: Photographs of a Man-Altered Landscape*[6] (1975), organized by William Jenkins, and from there to the aesthetic of August Sander and Walker Evans. Like Sander and Evans, the photographers featured in *New Topographics* preferred a detached style, which, in their case, was used to distance themselves from earlier more romanticized landscape photography. Instead of the monumental "wilderness" landscapes with overwhelming mountains and skies that inspired Ansel Adams (1902-1984) and others, the new topographers expressed their social and environmental concerns through images of the modified land in a muted, seemingly detached way. In Durden's (2003) introduction to *Hidden*, Seawright's photographs are linked to the quiet Holocaust photographs of Dirk Reinhartz and Mikael Levin as discussed in Ulrich Baer's *Spectral Evidence* (2002). According to Baer, Reinhartz and Levin show us "that there is something in a catastrophe as vast as the Holocaust that remains unassimilable" (pp. 66-67). Baer's work is concerned with the question of "how to frame an image with references to a historical event that consisted in the lasting destruction of most, and possibly all, explanatory referential frames and contexts for understanding" (p. 69). Seawright's photographs answer this question of the representation of extreme violence in the negative, by focusing on the void, the placelessness of the Afghan setting. By providing no content and no spatial and temporal context, Seawright's photographs suggest that war is entirely meaningless. The photographs convey a world that is both spatially and temporally disconnected, and they offer no psychological solace. His work addresses the issue of the inability to access the war zone, to render war visible at any given time, to give it any meaning at all. But for Baer, the empty Holocaust landscapes are more than expressions of "the unspeakable" (p. 69). They are deemed important as disruptions of the flood of Holocaust references in the cultural and entertainment industries; they are also seen as a critical commentary on the appropriation of landscape art by the Nazis, and as evidence of the cover-up of their activities. Here the comparison between Reinhartz/Levin and Seawright breaks down. In the case of Afghanistan, the public has never faced a flood of historical and contextual explanations. On the contrary, there appears to be a general *lack* of explanations. The classic ethnography of Afghanistan, that of Dupree (1973), is more than 30 years old, and media interest in the country has been erratic at best. By 2006, Sherry Ricchiardi (2006) of the *American Journalism Review* concluded that Afghanistan had become a "forgotten war." Two years later, in another piece for *American Journalism Review*, she found that little had changed (Ricchiardi, 2008). Joshua Foust (2008) of the *Columbia Journalism Review* similarly agreed that the media produced "a badly incomplete picture of what Afghanistan is really like" (para 8). In 2008, U.S. media spent an average 0.9% of news holes on Afghanistan, according to the Pew Project for Excellence in Journalism (2009). Although media interest in Afghanistan was greater at the time that Seawright made *Hidden*, it was far from comparable to the decades of Holocaust narrativization. In the case of Afghanistan, not the presence but the absence of contextual explanations seems to be the problem. I find it therefore difficult to read *Hidden* entirely like the empty Holocaust landscapes. Although the photographs in *Hidden* are powerful symbols of the destruction of war and invite the viewer to contemplate the limits of representation, the minimalist approach seems less effective as a commentary on contemporary discourse about Afghanistan (or rather the lack of any serious discourse). I do not mean to criticize either Seawright or Durden here but want to argue that, in a general sense, in the case of Afghanistan, we probably need *more not less* contextualization. The photographs in *Hidden* appear disconnected from the situational contexts in which they were produced. This turns them into effective and powerful symbols of war and violence but makes them less effective as a political commentary on the specifics of the war in Afghanistan. I do not think the photographer ever wanted to make or felt able to make such a commentary. Seawright's non-narrative work seems to be, as Durden noted, about the powerlessness of the observer, the inability to tell the others' story and to make their experience of war visible.

A minimalist strategy to cover conflict has been employed by others too, most notably by photographer Alfredo Jaar, to whom Durden refers in his introduction to *Hidden* as well. In *Real Pictures* (1995), part of a larger series called *The Rwanda Project* (1994-2000), Jaar buried images of the Rwanda genocide in boxes that looked like coffins and monuments for the dead. Printed texts on their tops described the inside and encouraged visitors to think about the victims of the genocide and to imagine the unimaginable atrocities they suffered. Jaar used the caption as a political device, to counter photography's tendency to beautify the world, in a way suggested by Benjamin (1982): "What we must demand from the photographer is the ability to put such a caption beneath his picture as will rescue it from the ravages of modishness and confer upon it a revolutionary use value" (p. 24). Jaar took this advice to the extreme by removing the photographic frame entirely and giving precedence to the text. In *Hidden*, the tendency in photography to aestheticize the world is not countered by assigning texts to the images. On the contrary, the viewer/reader is barely offered any textual clues at all. The minimal captions in *Hidden* follow conventional modern art practice and suggest a timeless universal quality, largely at the expense of the specifics of the world in which the images were made. The minimalist captions in *Hidden* make the photographs stand out as autonomous works of art.

The photographs are framed as art in two other ways in Durden's introduction: by referencing other works of art and by explicitly opposing *Hidden* to photojournalism. The introduction positions Seawright's work in opposition to "the trappings of the exoticising vision, best typified by the photojournalists' portrait of Afghanistan as a spectacle of ruins" and "the spectacle associated with contemporary media representations" (introduction, para 1, 3). It locates *Hidden* within the history of photography by referencing Alfredo Jaar, Robert Misrach, Seawright's own "Sectarian Murder" series (1988)[7] and the Holocaust photographs of Dirk Reinhartz and Mikael Levin as discussed in Ulrich Baer's *Spectral Evidence* (2002). It further positions it against Norfolk's work, with its focus on ruins and "visually sumptuous and spectacular" style (introduction, para 10).

The work itself offers the connoisseur-viewer the pleasure of recognition—one of Seawright's images looks surprisingly like Fenton's iconic photograph, "Valley of the Shadow of Death" (1855). *Hidden* ultimately appears to be more about photography and the photographer's position than about Afghanistan. The explicit framing of *Hidden* in art shifts attention from the photographed to the photographer. The work ceases to be "about Afghanistan" and becomes "about Seawright." As Crimp (1989) wrote about the transfer of images from various departments to a separate Art, Prints, and Photographs Division in the New York Public Library: "Thus, what was once housed in the Jewish Division under the classification 'Jerusalem' will eventually be found in Art, Prints, and Photographs under the classification 'Auguste Salzmann.' What was Egypt will become Beato, or du Camp, or Frith; Pre-Columbian Middle America will be Désiré Charnay; the American Civil War, Alexander Gardner and Timothy O'Sullivan; the cathedrals of France will be Henri LeSecq; the Swiss Alps, the Bisson Frères; the horse in motion is now Muybridge; the flight of birds, Marey; and the expression of emotions forgets Darwin to become Guillaume Duchêne de Boulogne" (p. 7). By a similar logic, the textual framing of *Hidden*, and of *Chronotopia* too, turns "Afghanistan" into "Norfolk and

Seawright." The photographs are framed as forms of autonomous art in such a way that they encourage the viewer to contemplate photography and the photographer rather than Afghanistan.

## THE PHOTO STUDIO AS HETEROTOPIA

Professional photographers have tried to give local agents a voice and to reveal realities not accessible to themselves by collecting and publishing "vernacular" material. Recent work based on such an approach includes Geert van Kesteren's *Baghdad Calling* (2008), with mobile phone snapshots taken by Iraqis, and Susan Meiselas' *Kurdistan* (2008), which also mixes amateur and professional photography. *Taliban* (2003) by Dworzak fits into this tradition.[8] The book opens with five of Dworzak's own images. The first four are in black and white. These are images of signs and ads with heads of people and mammals covered or removed, demonstrating the impact of Taliban politics on bodily appearance. The fifth image is in color and pictures a man with fashionable sandals and a rifle beautified with red tape. It hints at the studio portraits that follow and that Dworzak discovered when photo studios reopened in Kandahar in 2002.[9] As the book informs the reader, the portraits are mostly of Taliban members, who had them made in November 2001 when they visited the photo studio for passport pictures.

The book gives the images public, historical, political and aesthetic meanings that their makers certainly did not intend to make. But it also retains some of the intimate qualities of a private record. Its format (8.2 x 6.1 x 0.7 inches) is relatively small, and the images are not much bigger than the snapshots people paste in family albums. Although the composition and the pose of the depicted Taliban refer to the more formal approach of studio portraiture, the small white border around them is reminiscent of the family snapshot.

The portraits are of one, two, or three men. Some appear to be of comrades. Other photographs look like portraits of brothers, and one portrait seems to be of a father and his two sons. The double and triple portraits show us intimate relations. Some of the men are holding hands; others have their arms around the shoulder of their co-sitter. These members of the Taliban are not the antimodernists and savage warriors that the mainstream media made of the them: Some have their eyes lined with kohl, some pose with flowers, and others carry a modern wristwatch or sunglasses, or display a lock of hair. The men seem to appreciate the modern, the fashionable, the frivolous, and the Western. The Swiss Alps chalets and gardens in the background evoke a romantic, bourgeois atmosphere and the flowers in the foreground suggest hominess. The Alps setting creates an ambivalent layer, locating the sitters between the West and the East, and between the foreign and the domestic.

There are two main types of portraits in the book: medium close-ups against an empty background and portraits, from the waist up or full figure, against a backdrop with a photo or several photos from the Alps. The first were taken as large-format black-and-white pictures and later hand-colored. The second are simple color photos, developed by 1-hour services in Pakistan. Some of the props in these pictures (beads, a cigarette, flowers, a telephone), the backdrops (Alps chalets), and the way the Taliban relate to each other in some of the pictures suggest familial closeness. The depicted Taliban look in the lens, which transposes the intimacy of family relations onto the viewer. But there are also props that do not match the intimate frame: guns and rifles. The coexistence of these attributes referencing the homely and the violent creates an ambivalence that makes these images powerful (see Fig. 4.2).

**Fig. 4.2. Taliban (2002).** © Thomas Dworzak/Magnum Photos.

The men in *Taliban* appear as hybrid characters that resist easy categorization: homely and violent, private and public, local and global, traditional and modern, male and feminine. Perhaps the texts by Thomas Dworzak, John Lee Anderson, and Thomas Rees in the book remove some of the complexities that are thus evoked. They situate the portraits in their Afghan context, that of the city of Kandahar and its reputation as a city of homosexuality, pederasty, music, and play. Although this may explain the preference for make-up and decorum in the pictures, it does not entirely explain the ambivalence created by the co-presence of Swiss chalets, Kalashnikovs, plastic flowers, wristwatches, and telephones. The photo studio where these images were made apparently operated as a locus to stage a relationship between an idealized home, symbolized by the unreal flowers and the Swiss chalets poster, and an equally virtual violent culture, symbolized by the gun. The familiar attributes of homeliness and violence, used separately, would crudely categorize and stereotype but combined in a single space they create a novel unsettling experience. The *Taliban* pictures derive their power from the incompatible simultaneity of utopian and dystopian referencing. This brings to mind aspects of what Foucault (1984) called heterotopian space, which I understand as a marginal space that, through its combination of different other spaces, disrupts normal space and time. The photo studio was certainly marginal under the Taliban regime. It was tolerated for its surveillance capabilities—the production of passport pictures to turn subjects into state objects. But the *Taliban* portraits demonstrate that it also functioned as a liminal space for the enactment of deviant behavior. It provided the Taliban an entry into a different world that allowed them to temporarily acquire a new identity outside the realms of state surveillance. The studio constituted a break with both the normal space of the regime and the normal time of lived experience. According to Dworzak (2003), none of the photographers seemed to think of the pictures as contradictory, hypocritical or dissident. The Taliban regime is not contested through pure *opposition*. But its subversion is located in the heterotopian connection of hominess and violence. Perhaps the depicted Taliban were not fully aware of their subversion but it is difficult not to think of their pictures as forms of private resistance to the codes of the regime. At the same time, the pictures subvert the stereotypical view of the Taliban that emerged in the media after these portraits were made.[10] Originally intended for private use, the images in the book call into question the Western mainstream view of the Taliban as savage warriors and reveal to the viewer how local Taliban apparently resisted the regulation of their bodies by the state.

## THE OTHERNESS OF WAR

The photo-books discussed in this chapter were included in order to expand the range of frames found in the western European press and in U.S. public affairs. Haviv's book reproduces most of the common themes found there and does not present an alternative perspective. Norfolk's photos in *Chronotopia* look quite different from news pictures, mainly because they were made with a large-format camera in early morning light. The ruined landscape itself is a conventional news theme that has often been used to frame war as ubiquitous and destructive. Norfolk photographed the ruins and debris of war in a picturesque style with the intent to counter easy consumption and to stimulate contemplation. Seawright counter-framed the drama and singular focus of news media in content and style, with calm and bleak images of desolate landscapes and abandoned interiors. Both photographers created an aesthetic experience through technical perfection. Their able use of color and composition makes their landscapes stand out as potent symbols of war and the human condition. Their aesthetic strategies, I think, are less effective as a commentary on the war *in Afghanistan*. Drawing attention to the powerlessness of the observer and the limits of representation, Seawright seems to admit that Afghanistan is a territory that he simply cannot access properly. All the private spaces in his photographs are empty, and any narrative about the Afghan conflict or confrontation with the histories of named individual Afghans is avoided. Without any narrative, the viewer is left with the war landscape as an inexplicable terrain. Seawright's work suggests that in the visual representation of the Afghanistan intervention, it is not just the country and its inhabitants that constitute the other but war itself. War is a foreign experience that outsiders cannot access, and the photo-textual representations of Afghanistan are ultimately evidence of our inability to render war experience visible and understandable. *Taliban* is the only one of the four books that grants the viewer some access to Afghan realities. It is a rare document of the other side, offering the viewer a glimpse into the privacy of Taliban, revealing ways in which they framed their bodies. *Taliban* conveys a sense of a real space, between the utopian quest for freedom of *The Road to Kabul* and the dystopian landscapes of Norfolk and Seawright.

## NOTES

1. Walker Evans, who rejected the label "documentary" himself, became recognized as an important photographic artist in and through a retrospective of his work at the Museum of Modern Art in 1971. The

exhibition Walker Evans & Company (2000) confirmed his status as an influential artist in the history of photography.

2. Cornell Capa (born Kornel Friedmann, 1918-2008) used the term *concerned photographer* for a 1967 exhibition at the Riverside Museum in New York City with work by his brother Robert Capa, David Seymour, Werner Bischof, Dan Weiner, and Leonard Freed.
3. For a selection of images from *Afghanistan: Chronotopia* (Norfolk, 2002), see http://www.simonnorfolk.com.
4. For a selection of images from *Hidden* (Seawright, 2003), see http://www.paulseawright.com.
5. Norfolk referred to it in a piece called "Et in Arcadia ego" (*Even in Arcady I* [*death*] *am*). The phrase has been interpreted as a reference to the ambivalent experience of beauty and horror (Panofsky, 1983).
6. The exhibition showed work by Robert Adams, Lewis Baltz, Bernd and Hilla Becher, Joe Deal, Frank Gohlke, Nicholas Nixon, John Schott, Stephen Shore, and Henry Wessel, Jr.
7. "The Valley of Death" is a popular reference in photography. It also was used for an article on Afghanistan's Korengal Valley (Junger, 2008) with photographs by Tim Hetherington.
8. For a selection of images from *Taliban* (Dworzak, 2003), see http://www.magnumphotos.com.
9. Photography of living beings was forbidden under the Taliban but passport photography had been allowed at one point. When they had their passport pictures taken, some Taliban would ask for an additional portrait.
10. *De Morgen* (January 19, 2002) published several of the photographs from Dworzak's *Taliban* (2003).

# Conclusion

> As members of society, most of us see only what we expect to see, and what we expect to see is what we are conditioned to see when we have learned the definitions and classifications of our culture.
> —Victor W. Turner (1967)

When the U.S. and its allies decided to enter Afghanistan in 2001, they deployed a double discourse of militarism and humanitarianism to legitimate their intervention. The discourse of militarism was grounded in a framing of Afghanistan as a country of fundamentalists and terrorists, whereas the discourse of humanitarianism was construed around the victimized Afghan woman. The plot line was simple and dramatic: It was proposed that the West needed to defend itself against the terrorists, and that the Afghan population had to be saved from the fundamentalists. Photo-texts in the western European press predominantly reproduced this double discourse and its basic structure of problem, cause, treatment, and morality.

Afghanistan often was presented in the press, consistent with 19th-century colonial views, as an anachronistic space of savage warriors, and victimized women and children. It was framed through a projection in time as the Stone Age or the Middle Ages, words that were literally and frequently used by officials as well as journalists. Visually, the press supported the idea of a prehistoric or medieval Afghanistan through images of poverty, ruins, men on horseback, and traditional games and dresses. Afghan women and the Afghan landscape were represented as passive, isolated, and violated. Images of women in burqa symbolized their patriarchal and fundamentalist oppression. Beards, turbans, and weapons were singled out as attributive evidence of the combination of violence and religion. Differences among Afghans, when reported, were rarely explained as specific social, political, and ecological adaptations but attributed to a natural disposition toward disagreement and fighting. The press portrayed and explained the conflict in Afghanistan through an emphasis on the violent disposition of the other, and sometimes more generally suggested, in a Hobbesian way, that conflict was the natural condition of humankind. This created a view of war as *ubiquitous*.

This *ubiquity of war* frame was one of several coexisting frames that defined the nature of the conflict in Afghanistan in the first months of the intervention. The press also described the conflict in Afghanistan as a *clash of civilizations*. A bipolar view of the world was construed through photo-texts of a premodern and irrational Afghanistan on the one hand, and modern rational Western forces on the other hand. Representations of disciplined Western techno-warriors were contrasted with images of Afghan warlords, and peaceful demonstrations in the West with "violent" protests in the East. These images supported the treatment recommendation that a Western military intervention was needed unless the world wanted to be exposed to the cruelty of Afghanistan.

At the onset of the intervention, many press representations conveyed a strong belief in the Western capability to manage such an endeavor successfully. The press adopted the *managerial* frame, a major characteristic of DoD public affairs communication, and focused on military logistics, tactics, and strategies. This generally promoted a view of war as a technical operation that can be managed effectively rather than as *the failure to manage* disputes nonviolently. Two frames in official communication defined important conditions for the successful management of war: *unity* and *leadership*. The press only marginally adopted these frames. With the exception of initial reporting in the *Times*, the western European press did not encourage a rallying around the flag with images of familial, military, national, and allied homogeny. It usually preferred to frame the intervention in Afghanistan as a *U.S. and international* affair, a view reinforced by the occasional image of American patriotism. The western European press did domesticate the news, however, with photographs of national leaders, border controls at home, home-front scenes, references to national history, stories of one's own soldiers and national debates about troops and gear. Images of national leadership varied across the western European press but, again with the exception of the *Times*, national leaders were generally not presented as war leaders. This observation does not imply that the press was generally not in favor of the war. The newspapers certainly criticized the military strategy of the United States but the need for a military presence in Afghanistan was not seriously challenged in the first months of the intervention. Peace options were hardly explored and peace was largely approached from a dichotomous perspective as the *absence of war*. Well before U.S. President Barack Obama used the phrase "war of necessity" to describe the conflict in Afghanistan, the western European press generally accepted the idea that the West could not but intervene. Initially, the *Times* presented the intervention even as a moral duty.

The official double discourse of militarism and humanitarianism, which encouraged journalists to present the conflict in Afghanistan as a simple and unified story with villains, victims, and heroes, was largely adopted. But ambivalences about the war entered the mainstream media as well. Photo-texts in the press represented the conflict in Afghanistan as humanitarian *and* destructive. The *humanitarian* frame suggested that something good could come out of the military violence. It legitimated the intervention and countered anxieties by offering viewers the prospect of peace, progress, and prosperity. The *destruction* frame countered this idea. However, photo-texts of damage and injury did not usually imply a condemnation of the war, only of the war strategy.

When Kabul fell to the Northern Alliance, the western European press took a consciously ambivalent position. The capital was portrayed as a liminal zone, between death and resurrection. Simplified bipolar representations dissolved on various other occasions, in photo-texts that conveyed a sense of loss of control, a fear of disorder, and the convergence and obliteration of identities. The Western Taliban, the European Muslim, the Afghan refugee, the American *buzkashi* player, and the global terrorist were represented in ways that suggested a breakdown of primordial unities. Sometimes the boundaries between the conflict zone and the rest of the world dissolved too, for example, when the war appeared as a *hyper-reality* and photo-texts directed attention from the armed war to the image war. It seems that in times of war press photographs do not merely create a sense of cohesiveness and unification but express anxieties associated with the breakdown of imagined communities.

What largely remained absent from the news photography of the Afghan conflict is the human condition of the war, the experience of the local population, the prolonged consequences of the war, and its structural and cultural roots. Peace journalism proposes a shift of framing in order to make these issues central to news discourse. It seeks to counter-frame the dominant war-oriented coverage of conflict, but the request to cover conflict from multiple perspectives runs counter to the routine practice of indexing elite political and military views and to the economic demand to deliver news stories quickly and in a dramatic format. Peace journalism provides journalists with ethical guidelines for producing news but it says relatively little about the wider context in which journalists operate. Professional constraints, market demands, access restrictions, and language and cultural barriers all make peace reporting difficult to accomplish. It should therefore come as no surprise that peace journalism has until now remained relatively marginal in mainstream media. In order for peace journalism to be successful, Hackett (2006) argued that the normative principles of peace journalism need to be complemented with "a strategy based on a theoretically-informed analysis of the governing logics of news production" (p. 2). Structural reforms in the field of journalism seem a prerequisite for the development of peace journalism. The multiperspectival view proposed by peace journalism is impossible to achieve when the national press continues to domesticate what is going on abroad or when journalists routinely follow the official, national view or fall back on stereotypes because they have no time and resources to prepare and investigate issues thoroughly.

A major issue for photojournalists remains how to approach the underlying "invisible" aspects of conflict that are considered so important in peace journalism. Initially, with access denied by both the United States and the Taliban, even the visible aspects of the intervention in Afghanistan were kept out of sight. The type of warfare conducted in the country, including high-altitude bombing and secret operations, made traditional photojournalistic coverage of the war virtually impossible. The introduction of embedding later led to more coverage but also to a more narrow military view, reinforcing the idea that a military solution is the only answer to Afghanistan's problems.

For western Europeans, war remains a foreign experience that cannot be accessed, and the photo-textual representations of Afghanistan are ultimately evidence of this inability to render war visible. Like the magistrate in Coetzee's *Waiting for the Barbarians*, to whom I referred at the beginning of this book, the West is torn between the desire to tell the story of war and the other, and the inability to bear witness. Although photographic images can never be a substitute for experience, except in the mind of the fetishist, they may nevertheless work as an aid for understanding the world. Is the retreat into the modest minimalism of Seawright the only possible answer? One obvious pragmatic—but always inevitably partial—solution to the problems of representing foreign conflict could be to devote more space to the voices and images of the others in war. One route to addressing the issues of stereotyping, ethnocentrism, and simply access could be rethinking collaboration between Afghans and foreign journalists. The former cannot only provide an in-depth perspective of their own culture but they can also suggest new ways of photographing and representing conflict. Afghan photographers and journalists can help replace the image of Afghanistan as an empty space, a medieval realm, and an oppressive zone, and they can help connect with the multiplicities of living in contemporary Afghanistan. In too many instances, war remains a soldier's story that overexposes foreign troop movements, military gear, and strategies and tactics, whereas the social, cultural, political, and historical complexities of the conflict remain underexposed.

# References

Abu-Lughod, L. (2002). Do Muslim women really need saving? Anthropological reflections on cultural relativism and its Others. *American Anthropologist, 104*(3), 783-790.

AES+F. (1996-). Islamic project [installation and performance].

After September 11: Images from Ground Zero [Series of exhibitions]. (2002-2003). Washington, DC: State Department of the United States.

Ali, T. (2002). *The clash of fundamentalisms: Crusades, jihads and modernity*. London: Verso.

Allan, N. J. R. (2003). Rethinking governance in Afghanistan. *Journal of International Affairs, 56*(1), 193-202.

Althusser, L. (1971). *Lenin and philosophy and other essays* (B. Brewster, Trans.). New York: Monthly Review Press.

America's Army [Computer game]. (2002). U.S. Department of the Army. (Online at http://www.americasarmy.com.)

Anderson, B. (1983). *Imagined communities: Reflections on the origin and spread of nationalism*. London: Verso.

Appadurai, A. (1996). *Modernity at large: Cultural dimensions of globalization*. Minneapolis: University of Minnesota Press.

Apple, R. W. (2001, October 31). A nation challenged: News analysis; a military quagmire remembered: Afghanistan as Vietnam. *The New York Times*. Retrieved March 10, 2009, from http://query.nytimes.com/gst/fullpage.html?res=9906E6DF1230F932A05753C1A9679C8B63.

Armanios, F. (2003). The Islamic Traditions of Wahhabism and Salafiyya [CRS Report for Congress RS21695]. Retrieved July 16, 2004, from http://assets.opencrs.com/rpts/RS21695_20080124.pdf.

Arnheim, R. (1954). *Art and visual perception: A psychology of the creative eye*. Berkeley: University of California Press.

Azoy, G. W. (2003). *Buzkashi: Game and power in Afghanistan*. Philadelphia: University of Pennsylvania Press. (Original work published 1982)

Babcock, B. (1973). *The reversible world: Symbolic inversion in art and society*. Ithaca, NY: Cornell University Press.

Baer, U. (2002). *Spectral evidence: The photography of trauma*. Cambridge, MA: The MIT Press.

Bakhtin, M. (1981). *The dialogic imagination: Four essays* (C. Emerson & M. Holquist, Trans.). Austin: University of Texas Press.

Barnhurst, K. G. (1994). *Seeing the newspaper*. New York: St. Martin's Press.

Barrett, A. W., & Barrington, L. W. (2005). Is a picture worth a thousand words? *Press/Politics, 10(*4), 98-113.

Barry, J., & Thomas, E. (2009, February 9). Obama's Vietnam. *Newsweek*, pp. 14-21.

Barthes, R. (1973). *Mythologies* (A. Lavers, Trans.). London: Jonathan Cape. (Original work published 1957)

Barthes, R. (1977). *Image-music-text: Essays*. New York: Hill & Wang. (Original work published 1961)

Barthes, R. (1980). Rhetoric of the image. In A. Trachtenberg (Ed.), *Classic essays on photography* (3rd printing ed., pp. 269-286). New Haven, CT: Leete's Island Books. (Original work published 1964)

Bateson, G. (1972). *Steps to an ecology of mind: Collected essays in anthropology, psychiatry, evolution and epistemology*. Chicago: University of Chicago Press.

Baudrillard, J. (1994). *Simulacra and simulation* (S. F. Glaser, Trans.). Ann Arbor: University of Michigan Press. (Original work published 1981)

Baudrillard, J. (1995). *The Gulf War did not take place* (P. Patton, Trans.). Bloomington: Indiana University Press. (Original work published 1991)

BBC News—World Edition. (2003). Outrage at "Old Europe" remarks. Retrieved January 24, 2008, from http://news.bbc.co.uk/2/hi/europe/2687403.stm.

Beasley, R., & Danesi, M. (2002). *Persuasive signs: The semiotics of advertising*. New York: Mouton de Gruyter.

Beck, U. (2005). War is peace: On post-national war. *Security Dialogue, 36*(1), 5-26.

Becker, J. (1982). Communication and peace: The empirical and theoretical relation between two categories in social sciences. *Journal of Peace Research, 19*(3), 227-240.

Becker, K. E. (2003). Photojournalism and the tabloid press. In L. Wells (Ed.), *The photography reader* (pp. 291-308). London: Routledge.

Benford, R. D. (1997). An insider's critique of the social movement framing perspective. *Sociological Inquiry, 67*(4), 409–430.

Benford, R. D., & Snow, D. A. (1988). Ideology, frame resonance, and participant mobilization. In B. Klandermans, H. K. Kriesi, & S. Tarrow (Eds.), *International social movement research* (Vol. 1, pp. 197-217). London: JAI.

Benford, R. D., & Snow, D. A. (2000). Framing processes and social movements: An overview and assessment. *Annual Review of Sociology, 26*(1), 611-639.

Benjamin, W. (1968). The work of art in the age of mechanical reproduction (H. Zohn, Trans.). In H. Arendt (Ed.), *Illuminations: Essays and reflections* (pp. 217-251). New York: Harcourt. (Original work published 1936)

Benjamin, W. (1980). A short history of photography (P. Patton, Trans.). In A. Trachtenberg (Ed.), *Classic essays on photography* (3rd printing ed., pp. 199-216). New Haven, CT: Leete's Island Books. (Original work published 1931)

Benjamin, W. (1982). The author as producer. In V. Burgin (Ed.), *Thinking photography* (pp. 15-31). London: Macmillan Press. (Original work published 1934)

Bennett, W. L. (1990). Toward a theory of press–state relations in the U.S. *Journal of Communication, 40*(2), 103-125.

Bennett, W. L. (1995). *News: The politics of illusion*. New York: Longman.

Bennett, W. L., & Livingston, S. (2003). Gatekeeping, indexing and live-event news: Is technology altering the construction of news? *Political Communication, 20*(4), 363-380.

Benson, R. (2002). The political/literary model of French journalism: Change and continuity in immigration news coverage, 1973-1991. *Journal of Contemporary European Studies, 10*(1), 49-70.

Berger, J. (1972). *Ways of seeing*. London: British Broadcasting Corporation and Penguin Books.

Berger, P. L., & Luckmann, T. (1966). *The social construction of reality: A treatise in the sociology of knowledge*. Garden City, NY: Doubleday.

Berman, N. (2008). *Homeland*. London: Trolley.

Bertelsmann Stiftung. (Ed.). (2010). *Culture and conflict in global perspective: The cultural dimensions of global conflicts 1945-2007*. Gütersloh: Verlag Bertelsmann Stiftung.

Bezner, L. C. (1999). *Photography and politics in America: From the New Deal into the Cold War*. Baltimore: Johns Hopkins University Press.

Bhabha, H. (1994). *The location of culture*. London: Routledge.

Bijker, W. E., Hughes, T. P., & Pinch, T. J. (1987). *The social construction of technological systems: New directions in the sociology and history of technology*. Cambridge, MA: MIT Press.

Bivera, J. (2002). *U.S. Marines raid a suspected Al Qaeda hideout in the Helmand Province of Afghanistan* [Photograph]. Retrieved May 25, 2007, from http://www.defendamerica.mil/photoessays/mar2002/p032702a2.html.

Bolton, R. (Ed.). (1989). *The contest of meaning: Critical histories of photography*. Cambridge, MA: MIT Press.

Bond, B. (1977). *Liddell Hart: A study of his military thought*. New Brunswick, NJ: Rutgers University Press.

Bourdieu, P. (1990). *Photography: A middle-brow art* (S. Whiteside, Trans.). Stanford, CA: Stanford University Press. (Original work published 1965)

Boyd-Barrett, O., & Rantanen, T. (Eds.). (1998). *The globalization of news*. London: Sage.

Braestrup, P. (1983). *Big story: How the American press and television reported and interpreted the crisis of Tet 1968 in Vietnam and Washington*. New Haven, CT: Yale University Press.

Bresheeth, H. (2006). Projecting trauma: War photography and the public sphere. *Third Text, 20*(1), 57-71.

Brodsky, A. E. (2003). *With all our strength: The revolutionary association of the women of Afghanistan*. New York: Routledge.

Broomberg, A., & Chanarin, O. (2008, March 17). Unconcerned but not indifferent. *Foto8*. Retrieved 5 March, from http://www.foto8.com/home/index.php?option=com_content&task=view&id=377&Itemid=216.

Brothers, C. (1997). *War and photography: A cultural history*. London: Routledge.

Burgin, V. (1982a). Photography, phantasy, function. In V. Burgin (Ed.), *Thinking photography* (pp. 117-216). London: Macmillan Press.

Burgin, V. (Ed.). (1982b). *Thinking photography*. London: Macmillan Press.

Burke, E. (1998). *A philosophical enquiry into the origin of our ideas of the sublime and the beautiful*. Oxford: Oxford University Press. (Original work published 1757)

Buruma, I. (2001, November 6). Afghanistan vs. Vietnam. *Guardian*. Retrieved February 9, 2009, from http://www.guardian.co.uk/world/2001/nov/06/afghanistan.terrorism4.

Bush, G. W. (2001). Address to a joint session of Congress and the American people. Retrieved September 27, 2004, from www.whitehouse.gov/news/releases/2001/09/20010920-8.html.

Carey, J. W. (1989). *Communication as culture: Essays on media and society*. Boston: Unwin Hyman.

Carlebach, M. L. (1992). *The origins of photojournalism in America*. Washington, DC: Smithsonian Institution Press.

Castells, M. (1997a). *The information age: Economy, society and culture. Vol. 1: The rise of the network society*. Oxford: Blackwell.

Castells, M. (1997b). *The information age: Economy, society and culture. Vol. 2: The power of identity*. Oxford: Blackwell.

Ceulemans, C. (2007). The moral equality of combatants. *Parameters*, Winter, 99-109.

Clausen, L. (2004). Localizing the global: "Domestication" processes in international news production. *Media, Culture & Society, 26*(1), 25-44.

Clausewitz, C. (1976). *On war* (M. Howard & P. Paret, Trans.). Princeton, NJ: Princeton University Press. (Original work published 1932)

CNN. (2001, October 9). *Agencies question Afghan aid drops*. Retrieved January 28, 2008, from http://archives.cnn.com/2001/WORLD/europe/10/09/gen.aid.agencies/index.html.

Coetzee, J. M. (1982). *Waiting for the barbarians*. Harmondsworth, UK: Penguin Books.

Cohen, A. A., Mark, R. L., Roeh, I., & Gurevitch, M. (1996). *Global newsrooms, local audiences: A study of the Eurovision News Exchange*. London: John Libbey.

Colombani, J. M. (2001, September 13). Nous sommes tous Américains. *Le Monde*, p. 1.

Committee to Protect Journalists. (2001, November 8). *Al-Jazeera reporter expelled by Northern Alliance*. Retrieved 5 February, 2009, from http://cpj.org/2001/11/aljazeera-reporter-expelled-by-northern-alliance.php.

Committee to Protect Journalists. (2002, March 26). *Attacks on the press 2001: Afghanistan*. Retrieved 13 April, 2003, from http://cpj.org/2002/03/attacks-on-the-press-2001-afghanistan.php.

Committee to Protect Journalists. (2006, June 21). *Author Suskind alleges Afghan bombing of Al-Jazeera was deliberate*. Retrieved February 9, 2009, from http://cpj.org/2006/06/author-suskind-alleges-afghan-bombing-of-aljazeera.php.

Complex. (2003). *One World* [Magazine cover]. Retrieved February 28, 2010, from http://content.complex.com/assets/galleries/7692/big_oneworld.jpg.

Conversi, D. (2007). "We are all equals!" Militarism, homogenization and "egalitarianism" in nationalist state-building (1789-1945). *Ethnic and Racial Studies, 31*(4), 1-29.

Cooper, R. (2002, April 7). Why we still need empires. *The Observer*. Retrieved March 16, 2009, from http://observer.guardian.co.uk/worldview/story/0,11581,680117,00.html

Cooper, S. D. (2003). Press controls in wartime: The legal, historical, and institutional context. *American Communication Journal, 6*(4). Retrieved from http://acjournal.org/holdings/vol6/iss4/articles/cooper.htm

Cosgrove, T. (2001). *Tomahawk cruise missile* [Photograph]. Retrieved February 19, 2008, from http://www.defenselink.mil/photos/newsphoto.aspx?newsphotoid=3651.

Crews, R. D., & Tarzi, A. (2008a). Introduction. In R. D. Crews & A. Tarzi (Eds.), *The Taliban and the crisis of Afghanistan* (pp. 1-58). Cambridge, MA: Harvard University Press.

Crews, R. D., & Tarzi, A. (Eds.). (2008b). *The Taliban and the crisis of Afghanistan*. Cambridge, MA: Harvard University Press.

Crimp, D. (1989). The museum's old/the library's new subject. In R. Bolton (Ed.), *The contest of meaning: Critical histories of photography* (Vol. 22, pp. 3-14). Cambridge, MA: MIT Press. (Original work published 1981)

Curnalia, R. M. L. (2005). A retrospective on early studies of propaganda and suggestions for reviving the paradigm. *The Review of Communication, 5*(4), 237-257.

D'Angelo, P. (2002). News framing as a multiparadigmatic research program: A response to Entman. *Journal of Communication, 52*(4), 870-888.

Dannin, R. (Ed.). (2002). *Arms against fury: Magnum photographers in Afghanistan*. London: Thames & Hudson.

Dausend, P., & Haselberger, S. (2001, September 13). Wir sind alle Amerikaner. *Die Welt*, p. 7.

Davis, D. (2009, July 17). Afghanistan: The right war fought in the wrong way. *The Independent*. Retrieved July 19, 2009, from http://www.independent.co.uk/opinion/commentators/david-davis-afghanistan-ndash-the-right-war-fought-in-the-wrong-way-1750035.html?FORM=ZZNR4

de Saussure, F. (1983). *Course in general linguistics*. London: Duckworth. (Original work published 1916)

Debord, G. (1977). *La société du spectacle*. Paris: Buchet-Chastel.

Delahaye, L. (2001a, November 26). The fall of the Taliban. *Newsweek*, pp. 22-29.

Delahaye, L. (2001b). *Taliban, 2001* [Artwork].

Department of Defense. (2001). *Defend America*. Photo-essays. September-October 2001. Retrieved March 16, 2009, from http://www.defendamerica.mil/photoessays/photoessay0901.html.

Department of Defense. (2002). Army Spc. Jeffrey E. Reen [Photograph]. Retrieved May 25, 2007, from http://131.84.1.60/photos/Mar2002/020327-D-0000M-001.html.

Department of Defense. (2003). Information Operations Roadmap. Retrieved June 12, 2007, from http://www.gwu.edu/~nsarchiv/NSAEBB/NSAEBB177/info_ops_roadmap.pdf.

Department of Defense. (2008). *DoD Directive 5122.5*. Retrieved from http://www.dtic.mil/whs/directives/corres/pdf/512205p.pdf.

Department of Homeland Security. (2008). *Terminology to define the terrorists: Recommendations from American Muslims*. Retrieved June 12, 2008, from http://www.investigativeproject.org/documents/misc/126.pdf.

Domke, D. (2004). *God willing? Political fundamentalism in the White House, the "War on Terror," and the echoing press*. London: Pluto Press.

Domke, D., Perlmutter, D. D., & Spratt, M. (2002). The primes of our times? An examination of the "power" of visual images. *Journalism, 3*(2), 131-159.

Dorronsoro, G. (2005). *Revolution unending: Afghanistan, 1979 to the present*. New York: Columbia University Press.

Douglas, M. (2002). *Purity and danger: An analysis of the concept of pollution and taboo*. London: Routledge. (Original work published 1966)

Duffield, M. (2001). *Global governance and the new wars: The merging of development and security*. London: Zed Books.

Duganne, E. (2007). Photography after the fact. In M. Reinhardt, H. Edwards, & E. Duganne (Eds.), *Beautiful suffering: Photography and the traffic in pain* (pp. 57-74). Chicago: Williams College Museum of Art.

Dupree, L. (1973). *Afghanistan*. Princeton, NJ: Princeton University Press.

Durden, M. (2003). The poetics of absence: Photography in the "aftermath" of war. In P. Seawright (Ed.), *Hidden*. London: The Imperial War Museum.

Dworzak, T. (2003). *Taliban*. London: Trolley.

Edwards, H. (2007). Cover to cover: The life cycle of an image in contemporary visual culture. In M. Reinhardt, H. Edwards, & E. Duganne (Eds.), *Beautiful suffering* (pp. 75-92). Chicago: University of Chicago Press.

El Guindi, F. (1999). *Veil: Modesty, privacy and resistance*. New York: Berg.

Elkins, J. (2003). *Visual studies: A skeptical introduction*. New York: Routledge.

Entman, R. M. (1991). Framing U.S. coverage of international news: Contrasts in narratives of the KAL and Iran Air incidents. *Journal of Communication, 41*(4), 6-27.

Entman, R. M. (1993). Framing: Toward clarification of a fractured paradigm. *Journal of Communication, 43*(3), 51-58.

Entman, R. M. (2004). *Projections of power: Framing news, public opinions, and U.S. foreign policy*. Chicago: University of Chicago Press.

Esser, F. (1998). Editorial structures and work principles in British and German newsrooms. *European Journal of Communication, 13*(3), 375-405.

Esser, F. (1999). Tabloidization of news: A comparative analysis of Anglo-American and German press journalism. *European Journal of Communication, 14*(3), 291-324.

Evans, H. (1978). *Pictures on a page: Photo-journalism, graphics and picture editing*. New York: Holt, Rinehart & Winston.

Fahmy, S. (2004). Picturing Afghan women: A content analysis of AP wire photographs during the Taliban regime and after the fall of the Taliban regime. *Gazette, 66*(2), 91-112.

Fahmy, S. (2005). Photojournalists' and photo editors' attitudes and perceptions: The visual coverage of 9/11 and the Afghan war. *Visual Communication Quarterly, 12*(3), 146-163.

Featherstone, M. (1995). *Undoing culture: Globalization, postmodernism and identity*. London: Routledge.

Fenton, R. (1855). *Valley of the Shadow of Death* [Photograph].

Fialka, J. J. (1991). *Hotel warriors: Covering the Gulf.* Washington, DC: The Woodrow Wilson Center Press.

Finnegan, C. A. (2003). *Picturing poverty: Print culture and FSA photographs.* Washington, DC: Smithsonian Books.

Firth, J. R. (1957). *Papers in linguistics, 1934-1951.* London: Oxford University Press.

Fisk, R. (2001, November 29). We are the war criminals now. *The Independent.* Retrieved March 10, 2009, from http://www.independent.co.uk/opinion/commentators/fisk/robert-fisk-we-are-the-war-criminals-now-618426.html.

Foucault, M. (1980). *Power/knowledge: Selected interviews and other writings 1972-1977.* New York: Pantheon Books.

Foucault, M. (1984). Des espaces autres. *Architecture, Mouvement, Continuité, 5*, 46-49.

Foust, J. (2008, December 1). The unbearable lightness of Kabul. *Columbia Journalism Review.* Retrieved February 23, 2009, from http://www.cjr.org/behind_the_news/the_unbearable_lightness_of_ka.php.

Fowler, C. (2007). *Chasing tales: Travel writing, journalism and the history of British ideas about Afghanistan.* Amsterdam: Rodopi.

Fox, J. R., & Park, B. (2006). The "I" of embedded reporting: An analysis of CNN coverage of the "Shock and Awe" campaign. *Journal of Broadcasting & Electronic Media, 50*(1), 36-51.

Fox TV News (2001, November 5). *Special Report with Brit Hume* [Television broadcast].

Frank, R. (1959). *The Americans.* New York: Grove Press.

Frankenheimer, J. M. (1970). *The Horsemen* [Motion picture]. United States of America: Columbia TriStar Motion Picture Group.

Franklin, B. (1997). *Newszak and news media.* London: Arnold.

Franklin, B. (2008). The future of newspapers. *Journalism Practice, 2*(3), 306-317.

Freund, G. (1980). *Photography and society.* London: Gordon Fraser. (Original work published 1974)

Friedman, H. A. (2002). Psychological operations in Afghanistan. Retrieved March 9, 2009, from http://www.psywarrior.com/herbafghan.html and http://www.psywarrior.com/Herbafghan02.html

Galtung, J. (2000). The task of peace journalism. *Ethical Perspectives, 7*(2-3), 162-167.

Galtung, J. (2002). Peace journalism: A challenge. In W. Kempf & H. Luostarinen (Eds.), *Journalism and the new world order. Vol. 2: Studying war and the media* (pp. 259-272). Göteborg: Nordicom.

Galtung, J. (2006). Peace journalism as an ethical challenge. *GMJ: Mediterranean Edition, 1*(2). Retrieved June 12, 2009, from http://globalmedia.emu.edu.tr/fall2006/Fall_2006_Issue2/1Johan_Galtungpdf.pdf.

Gamson, W. A. (1988). Political discourse and collective action. In B. Klandermans, H. Kriesi, & S. Tarrow (Eds.), *From structure to action: Comparing social movement research across cultures* (pp. 219-244). Greenwich: JAI Press.

Gamson, W. A., & Lasch, K. E. (1983). The political culture of social welfare policy. In S. E. Spiro & E. Yuchtman-Yaar (Eds.), *Evaluating the welfare state: Social and political perspectives* (pp. 397-415). Paris: Academic Press.

Gamson, W. A., & Modigliani, A. (1989). Media discourse and public opinion on nuclear power: A constructionist approach. *American Journal of Sociology, 95*(1), 1-37.

Gannon, K. (2004, March 22). Road rage: Marauding Taliban and drug-dealing warlords on the road to Kandahar. *The New Yorker,* pp. 40-48.

Garamone, J. (2001, September 20). Rumsfeld says this war is different. *American Forces Press Service.* Retrieved June 17, 2008, from http://www.defenselink.mil/news/newsarticle.aspx?id=44849.

Garcia, M. R., & Stark, P. (1991). *Eyes on the news.* St. Petersburg, FL: The Poynter Institute for Media Studies.

Gitlin, T. (1980). *The whole world is watching: Mass media in the making and the unmaking of the new left.* Berkeley: University of California Press.

GlobalSecurity.org. (n.d.). 4th Psychological Operations Group (Airborne). Retrieved January 12, 2005, from http://www.globalsecurity.org/military/agency/army/4psyopgp.htm.

Goffman, E. (1971). *Relations in public: Micro studies of the public order.* New York: Basic Books.

Goffman, E. (1974). *Frame analysis: An essay on the organization of experience.* Cambridge, MA: Harvard University Press.

Gombrich, E. H. (1960). *Art and illusion: A study in the psychology of pictorial representation.* New York: Pantheon Books.

Goodson, L. P. (2001). *Afghanistan's endless war: State failure, regional politics, and the rise of the Taliban.* Seattle: University of Washington Press.

Gramsci, A. (1971). *Selections from the prison notebooks of Antonio Gramsci.* London: Lawrence & Wishart.

Gregory, D. (2004). *The colonial present.* Malden, MA: Blackwell.

Grier, P. (2002, April). The combination that worked. *Air Force Magazine.* Retrieved February 10, 2008, from http://www.airforce-magazine.com/MagazineArchive/Documents/2002/April%202002/0402combo.pdf.

Gurevitch, M., Levy, M., & Roeh, I. (1991). The global newsroom: Convergences and diversities in the globalization of television news. In P. Dahlgren & C. Sparks (Eds.), *Communication and citizenship: Journalism and the public sphere in the new media age* (pp. 195-216). London: Routledge.

Hackett, R. A. (2006). Is peace journalism possible? Three frameworks for assessing structure and agency in news media. *Conflict & Communication Online, 5*(2), 1-13.

Haigh, R. W., Pfau, M., Danesi, J., Tallmon, R., Bunko, T., Nyberg, S. et al. (2006). A comparison of embedded and nonembedded print coverage of the U.S. invasion and occupation of Iraq. *The International Journal of Press/Politics, 11*(139).

Haley, J. (1963). *Strategies of psychotherapy*. New York: Grune & Stratton.

Hall, S. (1981). The determinations of news photographs. In S. Cohen & J. Young (Eds.), *The manufacture of news: Deviance, social problems and the mass media* (pp. 226-243). London: Constable.

Halliday, M. A. K. (1966). Some notes on "deep" grammar. *Journal of Linguistics, 2*(1), 57-67.

Halliday, M. A. K. (1978). *Language as social semiotic: The social interpretation of language and meaning*. London: Edward Arnold.

Halliday, M. A. K. (2004). *An introduction to functional grammar* (revised by C. M. I. M. Matthiessen, 3rd ed.). London: Hodder Arnold. (Original work published 1994)

Hallin, D. (1989). *The "uncensored war": The media and Vietnam*. Berkeley: University of California Press.

Hallin, D. (1997). The media and war. In J. Corner, P. Schlesinger, & R. Silverstone (Eds.), *International media research: A critical survey* (pp. 206-231). London & New York: Routledge.

Hallin, D., & Gitlin, T. (1993). Agony and ritual: The Gulf War as popular culture and as a television drama. *Political Communication, 10*(4), 411-424.

Hampshire, J. (2002). Aviation ordnancemen [Photograph]. Retrieved March 17, 2009, from http://www.defendamerica.mil/photoessays/apr2002/p040102a1.html.

Handel, M. I. (1986). Clausewitz in the age of technology. In M. I. Handel (Ed.), *Clausewitz and modern strategy* (pp. 51-92). London: Frank Cass.

Hanitzsch, T. (2004). Journalists as peacekeeping force? Peace journalism and mass communication theory. *Journalism Studies, 5*(4), 483-495.

Haraway, D. (1991). *Simians, cyborgs and women: The reinvention of nature*. New York: Routledge.

Hariman, R., & Lucaites, J. L. (2007). *No caption needed: Iconic photographs, public culture, and liberal democracy*. Chicago: University of Chicago Press.

Harris, R. (1983). *Gotcha! The media, the government and the Falklands crisis*. London: Faber & Faber.

Hart, P., & Ackerman, S. (2001, November/December). Patriotism & censorship: Some journalists are silenced, while others seem happy to silence themselves. Retrieved April 6, 2003, from http://www.fair.org/index.php?page=1089.

Haviv, R. (2002a). *Afghanistan: The road to Kabul*. Millbrook: de.MO.

Haviv, R. (2002b). Shooting digital for the first time. *The Digital Journalist.* Retrieved August, 2, 2005, from http://digitaljournalist.org/issue0208/rh_intro.htm.

Hensher, P. (2001, October 17). Let's be honest: we need to impose our imperial rule on Afghanistan. *The Independent.* Retrieved March 18, 2009, from http://www.independent.co.uk/opinion/commentators/philip-hensher/philip-hensher-lets-be-honest-we-need-to-impose-our-imperial-rule-on-afghanistan-631620.html.

Here is New York [Organization]. (2002). *Here is New York: A democracy of photographs.* New York: Scalo.

Herman, E. S., & Chomsky, N. (1988). *Manufacturing consent: The political economy of the mass media.* New York: Pantheon.

Herold, M. W. (2004-2007). Scenes of Afghanistan, Parts 1-8. Retrieved on May 10, 2007, from http://pubpages.unh.edu/~mwherold/dossier.htm.

Hickey, N. (2002, January/February). Access denied: Pentagon's war reporting rules are toughest ever. *Columbia Journalism Review.* Retrieved January 1, 2005, from http://archives.cjr.org/year/02/1/hickey.asp.

Hicks, W. (1952). *Words and pictures: An introduction to photojournalism.* New York: Harper.

Hirschkorn, P. (2003, October 29). New York reduces 9/11 death toll by 40. *CNN.* Retrieved December 17, 2009, from http://www.cnn.com/2003/US/Northeast/10/29/wtc.deaths/.

History [Exhibition]. (2003). New York: Ricco/Maresca gallery.

Hofstede, G. (2003). *Culture's consequences: Comparing values, behaviors, institutions, and organizations across nations* (2nd ed.). Thousand Oaks, CA: Sage.

Houser, D. E. (2001). Aviation ordnancemen on the USS Enterprise [Photograph]. Retrieved April 6, 2008, from http://www.defenselink.mil/photos/newsphoto.aspx?newsphotoid=3653.

Hughes, A., & Noble, A. (2003). *Phototextualities: Intersections of photography and narrative.* Albuquerque: University of New Mexico Press.

Huston, J. (1975). *The Man Who Would Be King* [Motion picture]. United States of America: Allied Artists Pictures Corporation.

Hutton, B. G. (1983). *High Road to China* [Motion picture]. United States of America: Warner Bros.

Ignatieff, M. (2002, July 28). Nation-building lite. *New York Times Magazine.* Retrieved July 8, 2009, from http://www.nytimes.com/2002/07/28/magazine/nation-building-lite.html.

Independent Panel on Canada's Future Role in Afghanistan. (2008). [Report]. Retrieved March 10, 2009, from http://dsp-psd.pwgsc.gc.ca/collection_2008/dfait-maeci/FR5-20-1-2008E.pdf.

International Crisis Group. (2008). *Afghanistan: The need for international resolve* (Asia report No. 145). Retrieved February 18, 2008, from http://www.crisisgroup.org/library/documents/asia/south_asia/145_afghanistan___the_need_for_international_resolve.pdf.

Iyengar, S. (1991). *Is anyone responsible?: How television frames political issues.* Chicago: University of Chicago Press.

Jaar, A. (1995). Real Pictures [Installation].

Ji, L. J., Peng, K., & Nisbett, R. E. (2000). Culture, control, and perception of relationships in the environment. *Journal of Personality and Social Psychology, 78*(5), 943-955.

Joint Chiefs of Staff. (2003). Doctrine for Joint Psychological Operations, Joint Pub 3-53. Retrieved February 27, 2008, from http://www.dtic.mil/doctrine/jel/new_pubs/jp3_53.pdf.

Josephi, B. (2005). Journalism in the global age. *Gazette, 67*(6), 575-590.

Juluri, V. (2005). Media wars in Gandhian perspective. *Peace Review: A Journal of Social Justice, 17*, 397-402.

Junger, S. (2008). Into the Valley of Death. *Vanity Fair*. Retrieved February 9, 2009, from http://www.vanityfair.com/politics/features/2008/01/afghanistan200801.

Junne, G., & Verkoren, W. (2004). *Postconflict development: Meeting new challenges*. Boulder, CO: Lynne Rienner.

Kahneman, D., & Tversky, A. (1984). Choices, values, and frames. *American Psychologist, 39*(4), 341-350.

Kempf, W. (1996). Konfliktberichterstattung zwischen Eskalation und Deeskalation. Ein sozialpsychologisches Modell. *Wissenschaft und Frieden*. Retrieved August 25, 2004, from http://www.uni-muenster.de/PeaCon/wuf/wf-96/9620304m.htm.

Kempf, W. (1997). Kriegspropaganda versus Friedens-Journalismus. Retrieved June 12, 2009, from http://globalmedia.emu.edu.tr/fall2006/Fall_2006_Issue2/1Johan_Galtungpdf.pdf.

Kempf, W. (1999, July). *Deescalation-oriented conflict coverage? The Northern Ireland and Israeli-Palestinian peace processes in the German press*. Paper presented at the IAMCR Scientific Conference at Leipzig, Germany. Retrieved March 10, 2009, from http://kops.ub.uni-konstanz.de/volltexte/1999/267/pdf/267_1.pdf.

Kennedy, L. (2003). Remembering September 11: Photography as cultural diplomacy. *International Affairs, 79*(2), 315-326.

Kienzle, T. (2001). *The Spirit of Berlin* [Photograph].

Kipling, R. (1966). *Kim*. London: Macmillan. (Original work published 1901)

Kipling, R. (1987). *The man who would be king and other stories*. Oxford: Oxford University Press. (Original work published 1888)

Klein, W. (1956). *Life is good and good for you in New York: Trance witness revels*. Paris: Editions du Seuil.

Knight, S. (2006, October 27). MoD bans ITV News from warzones. *Times Online*. Retrieved July 17, 2009, from http://www.timesonline.co.uk/tol/news/uk/article615727.ece.

Knightley, P. (2002). Losing friends and influencing people. *Index on Censorship, 31*(1), 146-155.

Kobré, K. (2008). *Photojournalism: The professional's approach* (6th ed.). Boston: Focal Press.

Koening, T. (2006). Compounding mixed-methods problems in frame analysis through comparative research. *Qualitative Research, 6*(1), 61-76.

Koetzle, H.-M. (2002). *Photo icons: The story behind the pictures, 1928-1991* (Vol. 2). Köln: Taschen.

Kolhatkar, S., & Ingalls, J. (2006). *Bleeding Afghanistan*. New York: Seven Stories Press.

Korea: The impact of war in photographs [Exhibition]. (1951). New York: Museum of Modern Art.

Korski, D. (2008). *Afghanistan: Europe's forgotten war* [Policy paper]. Retrieved June 12, 2009, from http://ecfr.3cdn.net/ fcdc73b8da7af 85936_q8m6b5o4j.pdf.

Kracauer, S. (1980). Photography. In A. Trachtenberg (Ed.), *Classic essays on photography* (pp. 245-268). New Haven, CT: Leete's Island Books. (Original work published 1927)

Kress, G., & van Leeuwen, T. (2003). *Reading images: The grammar of visual design*. London: Routledge. (Original work published 1996)

Kristeva, J. (1982). *Powers of horror: An essay on abjection* (L. S. Roudiez, Trans.). New York: Columbia University Press.

Kunder, J. (2006, March, 9). USAID assistance to Afghanistan. Statement before the House Committee on International Relations. Retrieved July 15, 2009, from http://www.usaid.gov/press/speeches/2006/ty060309.html.

Kurtz, H. (2001, October 31). CNN chief orders "Balance" in war news; reporters are told to remind viewers why U.S. is bombing. *The Washington Post*, p. C1.

Lacey, N. (1998). *Image and representation: Key concepts in media studies.* New York: Palgrave.

Lakoff, G. (2006). *Whose freedom? The battle over America's most important idea*. New York: Picador.

Lamb, C. J. (2004). Information operations as a core competency. *Joint Force Quarterly*. Retrieved July 14, 2008, from http://www.dtic.mil/doctrine/jel/jfq_pubs/1536.pdf.

Lamb, C. J. (2005). Review of psychological operations: Lessons learned from recent operational experience. Retrieved February, 27, 2008, from www.fas.org/irp/eprint/lamb.pdf.

Laviolette, D. A. (2001). The night launch of a Navy F-14 Tomcat. Retrieved May 25, 2007, from http://131.84.1.60/photos/Oct2001/011007-N-4533L-001.html.

Law, J., & Hassard, J. (1999). *Actor network theory and after*. Oxford: Blackwell.

Lee, F., Hallahan, M., & Herzog, T. (1996). Explaining real-life events: How culture and domain shape attributions. *Personality and Social Psychology Bulletin, 22*(7), 732-741.

Lee, S. T., & Maslog, C. C. (2005). War or peace journalism? Asian newspaper coverage of conflicts. *Journal of Communication, 55*(2), 311-329.

Lefebvre, M. (2007). The art of pointing: On Peirce, indexicality, and photographic images. In J. Elkins (Ed.), *Photography theory* (pp. 220-244). New York: Routledge.

Levi Strauss, D. (2003). *Between the eyes: Essays on photography and politics*. New York: Aperture.

Levin, D. (2003). Structure of news coverage of a peace process: A test of the indexing and zero-sum hypothesis. *Press/Politics, 8*(4), 27-53.

Levine, J. (2009, February 19). Turning the page: The news of Europe's newspapers. *Time*, pp. 26-28.

Lewis, J. (2002). Speaking of wars. *Television and New Media, 3*(2), 169-172.

Lieven, A. (2008, June 11). The dream of Afghan democracy is dead. *The Financial Times*. Retrieved March 10, 2009, from http://www.ft.com/cms/s/0/f25de8f4-37b1-11dd-aabb-0000779fd2ac.html?nclick_check=1.

Lister, M. (2007). A sack in the sand: Photography in the age of information. *Convergence: The International Journal of Research into New Media Technologies, 13*(3), 251-275.

Louw, E. P. (2003). The "War against Terrorism": A public relations challenge for the Pentagon. *Gazette, 65*(3), 211-230.

Lutz, C. A. (2002). Making war at home in the United States: Militarization and the current crisis. *American Anthropologist, 104(*3), 723-735.

Lynch, J. (1998). *The peace journalism option*. Taplow: Conflict & Peace Forums.

Lynch, J., & McGoldrick, A. (2005). *Peace journalism*. Stroud: Hawthorne Press.

Lynn, M., Shavitt, S., & Ostrom, T. (1985). Effects of pictures on the organization and recall of information. *Journal of Personality and Social Psychology, 49*(5), 1160-1168.

MacDonald, P. (1988). *Rambo III* [Motion Picture]. United States of America: TriStar Pictures.

Maguire, K., & Lines, A. (2005, November 22). Exclusive: Bush plot to bomb his Arab ally. *Daily Mirror*. Retrieved February 9, 2009, from http://www.mirror.co.uk/news/tm_objectid=16397937&method=full&siteid=94762&headline=exclusive--bush-plot-to-bomb-his-arab-ally-name_page.html.

Malinowski, B. (2002). *Coral gardens and their magic: The description of gardening*. London: Routledge. (Original work published 1935)

Maloney, S. M. (2007). Conceptualizing the war in Afghanistan: Perceptions from the front, 2001-2006. *Small Wars and Insurgencies, 18*(1), 27-44.

Marks, L. U. (2000). *The skin of film: Intercultural cinema, embodiment and the senses*. Durham, NC: Duke University Press.

Marsden, P. (1998). *The Taliban: War, religion and the new order in Afghanistan*. London: Zed Books.

Martinec, R., & Salway, A. (2005). A system for image-text relations in new (and old) media. *Visual Communication, 4*(3), 337-371.

McClintock, A. (1995). *Imperial leather: Race, gender and sexuality in the colonial contest*. London: Routledge.

McCombs, M., Llamas, J. P., Esteban, L.-E., & Rey, F. (1997). Candidate images in Spanish elections: Second-level agenda-setting effects. *Journalism and Mass Communication Quarterly, 74*(4), 703-717.

Meiselas, S. (2008). *Kurdistan. In the shadow of history* (2nd, revised ed.). Chicago: University of Chicago Press. (Original work published 1997)

Mendelson, A. L., & Thorson, E. (2004). How verbalizers and visualizers process the newspaper environment. *Journal of Communication, 54*(3), 474–491.

Menon, K. (2001, October 17). *Covering the new war*. Retrieved April 20, 2007, from http://cpj.org/reports/2001/10/frontline.php.

Messaris, P. (1994). *Visual literacy: Image, mind & reality*. Boulder, CO: Westview Press.

Messaris, P. (1997). *Visual persuasion: The role of images in advertising*. Thousand Oaks, CA: Sage.

Messaris, P., & Abraham, L. (2001). The role of images in framing news stories. In S. D. Reese, O. H. Gandy, & A. E. Grant (Eds.), *Framing public life: Perspectives on media and our understanding of the social world* (pp. 215-226). Mahwah, NJ: Erlbaum.

Messier, G. (2001). *The "Shooter"* [Photograph]. Retrieved March 17, 2009, from http://www.defendamerica.mil/photoessays/p101101a2.html.

Metz, C. (1977). *Le signifiant imaginaire: Psychanalyse et cinéma*. Paris: Union générale d'éditions.

Milton, S. C. (2003). *Family and friends* [Photograph]. Retrieved May 25, 2007, from http://www.defendamerica.mil/homecoming/may2003/h052903a3.html .

Mitchell, W. J. T. (1994a). *Landscape and power*. Chicago: The University of Chicago Press.

Mitchell, W. J. T. (1994b). *Picture theory: Essays on verbal and visual representation*. Chicago: University of Chicago Press.

Mitchell, W. J. T. (1995). Interdisciplinarity and visual culture. *Art Bulletin, 77*(4), 540-544.

Moeller, S. D. (1999). *Compassion fatigue: How the media sell disease, famine, war and death*. New York: Routledge.

Mollison, J. (2002). *Benetton advertising campaign 2003* [Photographs].

Morley, A. (2002). Foreword: Vital signs. In R. Haviv (Ed.), *Afghanistan: The road to Kabul*. Millbrook: de.MO.

Morrison, D., & Tumber, H. (1988). *Journalists at war: The dynamics of news reporting during the Falklands Conflict*. London: Sage.

Müller, J. (2006, October 25). Zoveel mogelijk lood het maïsveld inschieten. *NRC Handelsblad*. Retrieved November 12, 2006, from http://www.nrc.nl/binnenland/article1736089.ece/Zoveel_mogelijk_lood_het_maiuml_sveld_inschieten.

Mulvey, L. (1989). *Visual pleasure and narrative cinema: Visual and other pleasures*. Basingstoke: Macmillan. (Original work published 1975)

Narula, S. (2004). Psychological Operations (PSYOPs): A conceptual overview. *Strategic Analysis, 28*(1), 177-192.

National Counterterrorism Center. (2008). Words that work and words that don't: A guide for counterterrorism communication. Retrieved January 28, 2009, from http://www.investigativeproject.org/documents/misc/127.pdf.

National Security Strategy of the United States of America. (2002). Retrieved July 14, 2009, from http://www.globalsecurity.org/military/library/policy/national/nss-020920.pdf.

National Strategy for Combating Terrorism. (2003). Retrieved March 16, 2009, from http://www.globalsecurity.org/security/library/policy/national/counter_terrorism_strategy.pdf.

Naumann, K., Shalikashvili, J., Inge, P. A., Lanxade, J., van den Breemen, H., Bilski, B. et al. (2007). *Towards a grand strategy for an uncertain world: Renewing transatlantic partnership*. Retrieved June 15, 2008, from http://www.csis.org/media/csis/events/080110_grand_strategy.pdf.

Newton, J. H. (2001). *The burden of visual truth: The role of photojournalism in mediating reality*. Mahwah, NJ: Erlbaum.

Noelle-Karimi, C. (2002). History lessons: Women as the focus of conflict in Afghanistan. *The Women's Review of Books, 19*(7), 1-4.

Nojumi, N. (2002). *The rise of the Taliban in Afghanistan: Civil war and the future of the region*. New York: Palgrave.

Norfolk, S. (2002). *Afghanistan: Chronotopia*. Stockport: Dewi Lewis.

Out of the Shadows: A photographic exhibit of women in the new Afghan society [Exhibition]. (2004). New York: United Nations.

Paivio, A. (1986). *Mental representations: A dual coding approach*. New York: Oxford University Press.

Pan, Z., & Kosicki, G. M. (1993). Framing analysis: An approach to news discourse. *Political Communication, 10*, 55-76.

Panofsky, E. (1983). *Meaning in the visual arts*. Chicago: University of Chicago Press. (Original work published 1955)

Panzer, M. (2005). *Things as they are: Photojournalism in context since 1955*. London: Chris Boot.

Parr, M., & Badger, G. (2004). *The photobook: A history* (Vol. 1). London: Phaidon.

Paul, C., & Kim, J. J. (2004). *Reporters on the battlefield: The embedded press system in historical context*. Santa Monica, CA: RAND.

Paulson, K. A. (2001). *Politically correct speech*. Retrieved January 6, 2004, from http://www.freedomforum.org/templates/document.asp?documentID=15083.

Peirce, C. S. (1958). *Collected papers* (P. Weiss & C. Hartshone, Eds.). Cambridge, MA: Harvard University Press. (Original work published 1931)

Pelton, R. Y. (2002). House of war: The uprising at Mazar-e Sharif [Transcript, Aired by CNN October 26, 2002]. Retrieved May 29, 2009, from http://transcripts.cnn.com/TRANSCRIPTS/0210/26/cp.00.html.

Peress, G. (1984). *Telex Persan: In the name of revolution*. Paris: Contrejour.

Peress, G. (1996). Bosnia: Uncertain paths to peace. *The New York Times*. Retrieved May 28, 2008, from http://www.nytimes.com/specials/bosnia/.

Perlmutter, D. D. (1998). *Photojournalism and foreign policy: Icons of outrage in international crises*. New York: Praeger.

Peters, J. M. (2005). Het beeld nader bekeken. In L. Pauwels & J. M. Peters (Eds.), *Denken over beelden: Theorie en analyse van het beeld en de beeldcultuur*. Leuven: Acco.

PEW Project for Excellence in Journalism. (2009). The state of the news media: Annual report on American journalism. Retrieved July 8, 2009, from http://www.stateofthemedia.org/2009/index.htm.

Pfau, M., Haigh, M. M., Shannon, T., Tones, T., Mercurio, D., Williams, R. et al. (2008). The influence of television news depictions of the images of war on viewers. *Journal of Broadcasting & Electronic Media, 52*(2), 303-322.

Pfeffer, J. (2003). Paper bullets: An interview with Herbert A. Friedman. *Cabinet*. Retrieved January 23, 2006, from http://www.cabinet-magazine.org/issues/12/pefferFriedman.php.

Phillips, C. (1982, Autumn). The judgement seat of photography. *October, 22*, 27-63.

Power in the Pacific: Battle photographs of our navy in action on the sea and in the sky [Exhibition]. (1945). New York: Museum of Modern Art.

Price, V., Tewksbury, D., & Powers, E. (1997). Switching trains of thought: The impact of news frames on readers' cognitive responses. *Communication Research, 24*, 481-506.

Priest, D. (2009, January 23). Bush's "War" on terror comes to a sudden end. *Washington Post*. Retrieved March 16, 2009, from http://www.washington-post.com/wp-dyn/content/article/2009/01/22/AR2009012203929_pf.html.

Radiant World. (2001). *Bush uses Taliban play book?* [Photomontage]. Retrieved May 27, 2007, from http://www.bushspeaks.com/ home.asp?did=40.

Rantanen, P. (2005). Non-documentary burqa pictures on the internet: Ambivalence and the politics of representation. *International Journal of Cultural Studies, 8*(3), 329-351.

Rashid, A. (2000). *Taliban: Militant Islam, oil and fundamentalism in Central Asia*. New Haven, CT: Yale University Press.

Rashid, A. (2008). *Descent into chaos: How the war against Islamic extremism is being lost in Pakistan, Afghanistan and Central Asia*. London: Allen Lane.

Reese, S. D. (2001). Prologue—framing public life: A bridging model for media research. In S. D. Reese, O. H. Gandy, & A. E. Grant (Eds.),

*Framing public life: Perspectives on media and our understanding of the social world* (pp. 7-31). Mahwah, NJ: Erlbaum.

Rengger, N. (2002). On the just war tradition in the twenty-first century. *International Affairs, 78*(2), 353-363.

Reporters sans frontières. (2000, September). The Taliban and the media: A country with no news or pictures. Retrieved March 22, 2004, from http://www.omaid.com/english_section/in_the_press/talibanMedia_ReportersWithoutBorders.htm.

Reporters sans frontières. (2001, February 1). The Taliban and the media: A country with no news or pictures. Retrieved February 5, 2009, from http://www.unhchr.ch/Huridocda/Huridoca.nsf/TestFrame/8a9f2701af9c1572c1256a0a0032d6f6?Opendocument.

Reporters sans frontières. (2002, April 24). Pakistan annual report 2002. Retrieved January 27, 2004, from http://www.rsf.org/ article.php3?id_article=1447.

Revolutionary Association of the Women of Afghanistan. (n.d.). Afghanistan under the U.S. 'war on terror' [Photo Gallery]. Retrieved December 18, 2004, from http://www.rawa.org/s-photos.htm.

Ricchiardi, S. (2006, August/September). The forgotten war. *American Journalism Review*. Retrieved July 12, 2008, from http://www.ajr.org/Article.asp?id=4162.

Ricchiardi, S. (2008, October/November). Offscreen. *American Journalism Review*. Retrieved July 8, 2009, from http://www.ajr.org/article.asp?id=4602.

Riger, S. (2005). What's wrong with empowerment? *American Journal of Community Psychology, 21*(3), 279-292.

Rinehart, M. (1989). *Toward better concepts of peace. Working paper 89-14*. Retrieved 15 May, 2003, from http://www.colorado.edu/conflict/full_text_search/PeacePapers/89-14.htm.

Road to victory: A procession of photographs of the nation at war [Exhibition]. (1942). New York: Museum of Modern Art.

Roeder, G. (1993). *The censored war: American visual experience during Word War Two*. New Haven, CT: Yale University Press.

Rolston, B., & McLaughlin, G. (2004). All news is local: Covering the war in Iraq in Northern Ireland's daily newspapers. *Journalism Studies, 5*(2), 191-202.

Rosaldo, R. (1995). Foreword (C. L. Chiappari & S. L. López, Trans.). In N. G. Canclini (Ed.), *Hybrid cultures: Strategies for entering and leaving modernity* (pp. xi-xvii). Minneapolis: University of Minnesota Press

Rosie, M., MacInnes, J., Petersoo, P., Condor, S., & Kennedy, J. (2004). Nation speaking unto nation? Newspapers and national identity in the devolved UK. *The Sociological Review, 52*(4), 437-458.

Roy, O. (1995). *Afghanistan: From holy war to civil war*. Princeton, NJ: The Darwin Press.

Rubin, B. R. (1995). *The search for peace in Afghanistan: From buffer state to failed state*. New Haven, CT: Yale University Press.

Rubin, B. R. (1999). Afghanistan under the Taliban. *Current History, 98*(625), 79-91.

Rubin, B. R. (2002). *The fragmentation of Afghanistan: State formation and collapse in the international system* (2nd ed.). New Haven, CT: Yale University Press. (Original work published 1995)

Rubin, M. (2000). Afghanistan: As bad as its reputation? *Middle East Quarterly, 7*(3), 55-65.

Rummel, R. J. (1981). *Understanding conflict and war. Vol 5. The just peace*. Beverly Hills, CA: Sage.

Rumsfeld, D. H. (2001, September 27). A new kind of war. Retrieved February 11, 2003, from http://polylogos.org/chronique/rumsfeld.html.

Rupp, R. (2006). High hopes and limited prospects: Washington's security and nation-building aims in Afghanistan. *Cambridge Review of International Affairs, 19*(2), 285-298.

Said, E. W. (1978). *Orientalism*. New York: Pantheon Books.

Sammons, D. H. (2004). Psyop and the problem of measures of effectiveness (MOE) for the combatant commander. Retrieved February 28, 2008, from http://handle.dtic.mil/100.2/ADA425993.

Save the Children. (2004). Provincial reconstruction teams and humanitarian-military relations in Afghanistan. Retrieved October 20, 2005, from http://www.rusi.org/downloads/assets/Save_the_Children_UK_-_PRTs_and_Humanitarian-Military_Relations_in_Afghanistan_2004_09.pdf.

Scheufele, D. A. (1999). Framing as a theory of media effects. *Journal of Communication, 49*(1), 103-122.

Scott, C. (1999). *The spoken image: Photography and language*. London: Reaktion Books.

Seawright, P. (2003). *Hidden*. London: The Imperial War Museum.

Sekula, A. (1984). *Photography against the grain: Essays and photo works 1973-83*. Halifax, Nova Scotia: Press of the Nova Scotia College of Art and Design.

Semetko, H. A., & Valkenburg, P. M. (2000). Framing European politics: A content analysis of press and television news. *Journal of Communication, 50*(2), 93-109.

Shalom, S. R. (2009). Far from infinite justice: Just war theory and Operation Enduring Freedom. *Arizona Journal of International & Comparative Law, 26*(3), 623-697.

Shanker, T., & Schmitt, E. (2004, December 13). Pentagon weighs use of deception in a broad arena. *New York Times*. Retrieved March 12, 2005, from http://www.nytimes.com/2004/12/13/politics/13info.html.

Sharkey, J. (1993). When pictures drive foreign policy. *American Journalism Review, 15*(10), 14-19.

Shaw, M. (2002). Risk-transfer militarism, small massacres and the historic legitimacy of war. *International Relations, 16*(3), 343-359.

Shoemaker, P. J., & Reese, S. D. (1996). *Mediating the message: Theories of influences on mass media content* (2nd ed.). White Plains, NY: Longman.

Snow, D. A., & Benford, R. D. (1992). Master frames and cycles of protest. In A. D. Morris & C. McClung-Mueller (Eds.), *Frontiers in social movement theory*. New Haven, CT: Yale University Press.

Sontag, S. (1979). *On photography*. London: Penguin Books. (Original work published 1973)

Sontag, S. (2003). *Regarding the pain of others*. New York: Farrar, Strauss & Giroux.

Sreberny-Mohammadi, A. (1991). The global and the local in international communication. In J. Curran & M. Gurevitch (Eds.), *Mass media and society*. London: Edward Arnold.

Stevenson, R. W. (2005, August 4). President makes it clear: Phrase is "War on Terror." *The New York Times*. Retrieved September 12, 2005, from http://www.nytimes.com/2005/08/04/politics/04bush.html?scp=3&sq=war%20on%20terror&st=cse.

Suau, A. (2004). *Fear this*. New York: Aperture.

Sullivan, J. (Writer, Director). (2003). *Home of the brave. Land of the free*. G. A. Andreassen (Producer). Norway: Panzerfilm AS.

Sullivan, K., & Pincus, W. (2005, November 23). Papers says Bush talked of bombing Arab TV Network. *The Washington Post*. Retrieved June 24, 2008, from http://www.washingtonpost.com/wp-dyn/content/article/2005/11/22/AR2005112201784.html.

Suskind, R. (2006). *The one percent doctrine: Deep inside America's pursuit of its enemies since 9/11*. New York: Simon & Schuster.

Swanson, D. L., & Smith, L. D. (1993). War in the global village: A seven-country comparison of television news coverage of the beginning of the Gulf War. In R. E. Denton (Ed.), *The media and the Persian Gulf War* (pp. 165-196). Westport, CT: Praeger.

Synnott, A. (1987). Shame and glory: A sociology of hair. *The British Journal of Sociology, 38*(3), 381-413.

Szarkowski, J. (1966). *The photographer's eye*. New York: Museum of Modern Art.

Tagg, J. (1988). *The burden of representation: Essays on photographies and histories*. Hampshire, UK: Palgrave.

Tagg, J. (2009). *The disciplinary frame*. Minneapolis: The University of Minnesota Press.

Talbot, W. H. F. (1969). *The pencil of nature*. New York: Da Capo. (Original work published 1844)

Tankard, J. W. (2001). The empirical approach to the study of media framing. In S. D. Reese, O. H. Gandy, & A. E. Grant (Eds.), *Framing public life. Perspectives on media and our understanding of the social world*. Mahwah, NJ & London: Erlbaum.

Tarzi, A. (2008). The neo-Taliban. In R. D. Crews & A. Tarzi (Eds.), *The Taliban and the crisis of Afghanistan* (pp. 274-310). Cambridge, MA: Harvard University Press.

Tarzi, S. M. (1993). Afghanistan in 1992. *Asian Survey, 33*(2), 165-175.

Taylor, J. (1991). *War photography: Realism in the British press*. London: Routledge.

Tehranian, M. (2002). Peace journalism: Negotiating global media ethics. *Harvard International Journal of Press Politics, 7*(2), 58-83.

The Associated Press. (2001a, October 12). News media, administration struggle over press freedom, national security. Retrieved April 20, 2007, from http://www.freedomforum.org/templates/document.asp?documentID=15129.

The Associated Press. (2001b, November 9). Pentagon: U.S. troops in central Asia to remain off-limits to reporters. Retrieved April 20, 2007, from http://www.freedomforum.org/templates/document.asp?documentID=15326.

The Family of Man [Exhibition]. (1955). New York: Museum of Modern Art.

The Hague Centre for Strategic Studies. (2008). *Eyes wide shut? The impact of embedded journalism on Dutch newspaper coverage of Afghanistan*. The Hague.

The National Security Archive. (2006). *Rumsfeld's roadmap to propaganda*. National Security Archive Electronic Briefing Book No. 177. Retrieved 12 June, 2006, from http://www.gwu.edu/~nsarchiv/NSAEBB/NSAEBB177/index.htm.

Thomas, P. (2003). The anthrax attacks: A journalist assesses what went wrong in coverage of this story. *Nieman Reports* (Spring), 11-14.

Tomlin, J. (2002, August 16). CNN chief claims US media "censored" war. *Press Gazette*. Retrieved October 25, 2005, from http://www.pressgazette.co.uk/story.asp?sectioncode=1&storycode=21773.

Tuchman, G. (1978). *Making news: A study in the construction of reality*. New York: Free Press.

Turner, V. W. (1967). *The forest of symbols: Aspects of Ndembu ritual*. Ithaca, NY: Cornell University Press.

Turner, V. W. (1974). Liminal to liminoid in play, flow, and ritual: An essay in comparative symbology. *Rice University Studies, 60*(3), 53-92.

Turner, V. W. (1982). *From ritual to theatre: The human seriousness of play*. New York: Performing Arts Journal Publications.

Tyler, P. (2001, October 9). A nation challenged: The attack; Jets pound Taliban sites a 2nd night; F.B.I. agents shift antiterror tactics. *The New York Times*. Retrieved March 10, 2009, from http://query.nytimes.com/gst/fullpage.html?res=9D05EFDA113CF93AA35753C1A9679C8B63.

United Nations Development Program. (2007). *The state of human development and the Afghan Millennium Goals*. Retrieved February 8, 2008, from http://www.undp.org.af/MDGs/nhdr07_chap1.pdf.

United States Agency for International Development. (n.d.). Out of the shadows: Women in Afghan society [Online exhibit]. Retrieved October 28, 2005, from http://www.usaid.gov/missions/shadows/.

United States Agency for International Development. (2002a). Local Afghan girl [Photograph]. Retrieved May 18, 2009, from http://www.usaid.gov/locations/asia_near_east/countries/afghanistan/photogallery/gallery_05/photo12.html.

United States Agency for International Development. (2002b). USAID assists during Loya Jirga, June 2002 [Photograph]. Retrieved March 18, 2009, from http://www.usaid.gov/locations/asia_near_east/countries/afghanistan/photogallery/gallery_05/photo01.html.

United States Agency for International Development. (2003). Local Afghans observing the paving of the Kabul Kandahar road [Photograph]. Retrieved March 10, 2009, from http://gemini.info.usaid.gov/photos/displayimage.php?album=121&pos=11.

United States Agency for International Development. (2004). Out of the shadows: A photographic exhibit of women in the new Afghan society [Press release]. Retrieved March 18, 2009, from http:// www.usaid.gov/press/mediaadvisories/2004/ma040310.html.

United States Agency for International Development. (2005). USAID/Afghanistan Strategic Plan 2005-2010. Retrieved February 22, 2008, from http://www.usaid.gov/locations/asia_near_east/documents/countries/afghanistan/Afghanistan_2005-2010_Strategy.pdf.

United States Agency for International Development. (2004, October). Afghanistan Reborn [Magazine]. Retrieved October 28, 2005, from http://www.usaid.gov/locations/asia_near_east/documents/countries/afghanistan/AfghanistanReborn.pdf.

United States Army War College. (2006). Information operations primer: Fundamentals of Information Operations AY07. Retrieved June 18, 2007, from http://www.iwar.org.uk/iwar/resources/primer/info-ops-primer.pdf.

van Kesteren, G. (2008). *Baghdad calling*. Rotterdam: Episode.

Vogelsang, W. (2002). *The Afghans*. Oxford: Blackwell.

von Amelunxen, H., Iglhaut, S., & Rötzer, F. (Eds.). (1996). *Photography after photography: Memory and representation in the digital age*. München: G+B Arts.

Walker Evans & Company [Exhibition]. (2000). New York: Museum of Modern Art.

Walzer, M. (1977). *Just and unjust wars: A moral argument with historical illustrations*. New York: Basic Books.

Wasserman, E. (2001). The videophone war. *American Journalism Review*. Retrieved January 17, 2002, from http://www.ajr.org/Article.asp?id=2780.

Williams, B. A. (2003). The new media environment: Internet chatrooms and public discourse after 9/11. In D. K. Thussu & D. Freedman (Eds.),

*War and the media: Reporting conflict 24/7* (pp. 176-189). London: Sage.

Wolfsfeld, G. (1997). *Media and political conflict: News from the Middle East*. New York: Cambridge University Press.

Wolfsfeld, G. (2001). The news media and peace processes, the Middle East and Northern Ireland. *Peaceworks, 37*(2).

Worldviews [Organization]. (2002). Europeans see the world as Americans do, but critical of U.S. foreign policy [Report]. Retrieved September 10, 2006, from http://www.worldviews.org/key_findings/transatlantic_report.htm.

Wright, T. (2004). Collateral coverage: Media images of Afghan refugees, 2001. *Visual Studies, 19*(1), 97-112.

Zaller, J. (1996). The myth of massive media impact revived: New support for a discredited idea. In D. C. Mutz, P. M. Sniderman, & R. A. Brody (Eds.), *Political persuasion and attitude change* (pp. 17-78). Ann Arbor: University of Michigan Press.

Zeide, E. J. (2005). In bed with the military: First Amendment implications of embedded journalism. *New York University Law Review, 80* (October), 1309-1343.

Zelizer, B. (Ed.). (2001). *Visual culture and the Holocaust*. New Brunswick, NJ: Rutgers University Press.

Zelizer, B. (2004). When war is reduced to a photograph. In S. Allan & B. Zelizer (Eds.), *Reporting war* (pp. 115-135). London: Routledge.

Zillmann, D., Knobloch, S., & Yu, H.-S. (2001). Effects of photographs on the selective reading of news reports. *Media Psychology, 3*(4), 301-324.

# Author Index

# Subject Index

www.ingramcontent.com/pod-product-compliance
Lightning Source LLC
Jackson TN
JSHW071759020526
101819JS00001B/36
* 9 7 8 1 6 1 2 8 9 0 3 9 5 *